MATH ESSENTIALS

Math Essentials

Conquer Fractions, Decimals and Percentages— Get the Right Answer Every Time!

Second Edition

Steven Slavin

NEW YORK

Library of Congress Cataloging-in-Publcation Data

Slavin, Stephen L.
    Math essentials : conquer fractions, decimals, and percentages—get the right answer every time! / Steven Slavin—2nd ed.
        p. cm.
    ISBN 1-57685-305-5 (pbk.)
    1. Arithmetic. I. Title.

QA107 .S5384 2000
513.2—dc21   00-032709

Printed in the United States of America
9 8 7 6 5 4 3
Second Edition

For Further Information
For information on LearningExpress, other LearningExpress products, or bulk sales, please write to us at:
    LearningExpress™
    900 Broadway
    Suite 604
    New York, NY 10003

Or visit us at:
    www.learnatest.com

CONTENTS

INTRODUCTION

If you're like most other people, you use a pocket calculator to do your basic arithmetic. The calculator is fast and accurate as long, of course, as you punch in the right numbers. So what could be bad about a tool that saves you so much work and gives you the right answers?

Let me be brutally frank. You know why you bought this book, and it's not for the story. By working your way through this book, problem by problem, you will be amazed by how much your math skills will improve. But—and this is a really big BUT—I don't want you to use your calculator at all. So put it away for the time you spend working through this book. And who knows—you may never want to use it again.

Your brain has its own built-in calculator, and it, too, can work quickly and accurately. But you know the saying, "Use it or lose it."

The book is divided into four sections—a review of basic arithmetic, and then sections on fractions, decimals, and percentages. Each section is subdivided into four to eight lessons, which focus on building specific skills, such as converting fractions into decimals, or finding percentage changes. You'll then get to use these skills by solving word problems in the applications section. There are 21 lessons plus four review lessons, so if you spend 20 minutes a day working out the problems in each lesson, you can complete the entire book in about a month.

One thing that distinguishes this book from most other math books is that virtually every problem is followed by its full solution. I don't believe in skipping steps. You, of course, are free to skip as many steps as you wish, as long as you keep getting the right answers. Indeed, there may well be more than one way of doing a problem, but there's only one right answer.

When you've completed this book, you will have picked up some very useful skills. You can use these skills to figure out the effect of mortgage rate changes and understand the fluctuations in stock market prices or how much you'll save on items on sale at the supermarket. And you'll even be able to figure out just how much money you'll save on a low-interest auto loan.

Once you've mastered fractions, decimals, and percentages, you'll be prepared to tackle more advanced math, such as algebra, business math, and even statistics. At the end of the book, you'll find my list of recommended books to further the knowledge you gain from this book (see **Additional Resources**).

If you're just brushing up on fractions, decimals, and percentages, you probably will finish this book in less than 30 days. But if you're learning the material for the first time, then please take your time. And whenever necessary, repeat a lesson, or even an entire section. Just as Rome wasn't built in a day, you can't learn a good year's worth of math in just a few weeks.

While I'm doing clichés, I'd like to note that just as a building will crumble if it doesn't have a strong foundation, you can't learn more advanced mathematical concepts without mastering the basics. And it doesn't get any more basic than the concepts covered in this book. So put away that calculator, and let's get started.

SECTION 1

REVIEWING THE BASICS

On every page of this book you're going to be playing with numbers, so I want you to get used to them and be able to manipulate them. In this section you'll review the basic operations of arithmetic—addition, subtraction, multiplication, and division.

Your skills may have grown somewhat rusty. Or, as the saying goes, if you don't **use** it, you'll **lose** it. This section will quickly get you back up to speed. Of course different people work at different speeds, so when you're sure you have mastered a particular concept, feel free to skip the rest of that lesson and go directly to the next. On the other hand, if you're just not getting it, then you'll need to keep working out problems until you do.

Indeed, the basic way most students learn math is through repetition. It would be great if you could get everything right the first time. Of course if you could, then this book and every other math book would be a lot shorter. Once you get the basics down, there's no telling how far you'll go. So what are we waiting for? Let's begin.

PRETEST

The first thing you're going to do is take a short pretest to give you an idea of what you know and what you don't know. This pretest covers only addition, subtraction, multiplication, and division; all of which are necessary for learning the other concepts we will be studying later on in this book. Remember, you must not use a calculator. The solutions—completely worked out so you can see exactly how to do the problems—follow immediately after the pretest for you to check your work.

Add each of these columns of numbers.

1.	29		**3.**	1,025
	34			872
	16			2,097
	44			1,981
	37			655
	23			2,870
	56			+ 3,478
	+ 21			
			4.	19,063
2.	402			12,907
	199			10,184
	276			7,602
	850			14,860
	727			23,968
	233			17,187
	+ 196			+ 10,493

Do each of these subtraction problems.

5.	74		**7.**	2,436
	− 29			− 1,447
6.	335		**8.**	94,032
	− 286			− 76,196

Do each of these multiplication problems.

9.	49		**11.**	2,849
	× 96			× 7,491
10.	309		**12.**	56,382
	× 783			× 96,980

Do each of these division problems.

13. $7 \overline{)3,846}$ **16.** $29 \overline{)17,302}$

14. $9 \overline{)4,077}$ **17.** $79 \overline{)84,011}$

15. $6 \overline{)9,375}$ **18.** $364 \overline{)295,745}$

SOLUTIONS

1.
$$\overset{4}{29}$$
34
16
44
37
23
56
$+ \, 21$
260

2.
$$\overset{3\,3}{402}$$
199
276
850
727
233
$+ \, 196$
2,883

3.
$$\overset{3\;4\,2}{1,025}$$
872
2,097
1,981
655
2,870
$+ \, 3,478$
12,978

4.
$$\overset{3\;4\;4\;3}{19,063}$$
12,907
10,184
7,602
14,860
23,968
17,187
$+ \, 10,493$
116,264

5.
$$\overset{6\;1}{\cancel{7}4}$$
$- \, 29$
45

6.
$$\overset{2\;12\;1}{\cancel{33}5}$$
$- \, 286$
49

7.
$$\overset{1\;13\;12\;1}{2,4\cancel{3}6}$$
$- \, 1,447$
989

8.
$$\overset{8\;13\;9\;12\;1}{94,\cancel{0}32}$$
$- \, 76,196$
17,836

9.
$$
\begin{array}{r}
49 \\
\times\,96 \\
\hline
294 \\
4\,41 \\
\hline
4{,}704
\end{array}
$$

10.
$$
\begin{array}{r}
309 \\
\times\,783 \\
\hline
927 \\
24\,72 \\
216\,3 \\
\hline
241{,}947
\end{array}
$$

11.
$$
\begin{array}{r}
2{,}849 \\
\times\,7{,}491 \\
\hline
2\,849 \\
256\,41 \\
1\,139\,6 \\
19\,943 \\
\hline
21{,}341{,}859
\end{array}
$$

12.
$$
\begin{array}{r}
56{,}382 \\
\times\,96{,}980 \\
\hline
4\,510\,560 \\
50\,743\,8 \\
338\,292 \\
5\,074\,38 \\
\hline
5{,}467{,}926{,}360
\end{array}
$$

13.
$$
7\,)\overline{3{,}8\overset{36}{4}6}\;\;\;549\;\text{R3}
$$

14.
$$
9\,)\overline{4{,}0\overset{42}{7}7}\;\;\;453
$$

15.
$$
6\,)\overline{9{,}3\overset{3\,31}{7}5}\;\;\;1562\;\text{R3}
$$

16.
$$
\begin{array}{r}
596\;\text{R18} \\
29\,)\overline{1\overset{6\;1}{7}{,}302} \\
-14\;5\text{XX} \\
\hline
2\,\overset{71}{8}0 \\
-2\,61 \\
\hline
1\overset{8\,1}{9}2 \\
-174 \\
\hline
18
\end{array}
$$

17.

$$\begin{array}{r} 1{,}063 \text{ R}34 \\ 79 \overline{)\ 84{,}011} \end{array}$$

```
                1,063  R34
             71
       79 ) 84,011
          –79 XXX
            4 91
            5̶0̶1
            –4 74
              6 1
             2̶7̶1
              237
               34
```

18.

```
              812  R177
     364 ) 295,745
         –2912 XX
            3 1
           4̶ 54
           3 64
            8 9 1
           9̶0̶5
            728
            177
```

NEXT STEP

If you got all 18 problems right, then you probably can skip the rest of this section. Glance at the next four lessons, and, if you wish, work out a few more problems. Then go on to Section II.

  If you got any questions wrong in addition, subtraction, multiplication, or division, then you should definitely work your way through the corresponding lessons.

REVIEW LESSON 1

This lesson reviews how to add whole numbers. If you missed any of the addition questions in the pretest, this lesson will guide you through the basic addition concepts.

ADDITION

A ddition is simply the totaling of a column, or columns, of numbers. Addition answers the following question: How much is this number plus this number plus this number?

ADDING ONE COLUMN OF NUMBERS

Start by adding the following column of numbers.

Problem:
$$
\begin{array}{r}
6 \\
8 \\
4 \\
3 \\
5 \\
2 \\
+\,9 \\
\hline
\end{array}
$$

Did you get 37? Good. A trick that will help you add a little faster is to look for combinations of tens. Tens are easy to add. Everyone knows 10, 20, 30, 40. Look back at the problem you just did, and try to find sets of two or three numbers that add to ten.

What did you find? I found the following sets of ten.

Solution:
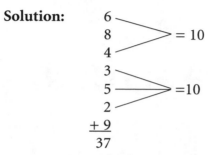

$$
\begin{array}{r}
6 \\
8 \\
4 \\
3 \\
5 \\
2 \\
+\,9 \\
\hline
37
\end{array}
$$

Here's another column to add. Again, see if you can find sets of tens.

Problem:
$$
\begin{array}{r}
3 \\
8 \\
2 \\
5 \\
7 \\
4 \\
1 \\
8 \\
6 \\
3 \\
4 \\
8 \\
+\,4 \\
\hline
\end{array}
$$

Did you get 63? I certainly hope so. Look at the tens marked in the solution.

Solution:

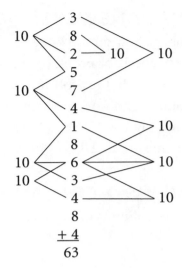

```
                3
        10      8
                2 ──► 10 ──► 10
                5
        10      7
                4
                  1 ──► 10
                8
        10      6 ──► 10
        10      3
                4 ─────► 10
                8
              + 4
               63
```

As you can see, there are a lot of possibilities, some of them over-lapping. Do you have to look for tens when you do addition? No, certainly not. But nearly everyone who works with numbers does this automatically.

ADDING TWO COLUMNS OF NUMBERS

Now let's add two columns of numbers.

Problem:
```
    24
    63
    43
    18
    52
  + 70
```

I'll bet you got 270. You carried a 2 into the second column because the first column totaled 20.

Solution:
```
   2
    24
    63
    43
    18
    52
  + 70
   270
```

HOW TO CHECK YOUR ANSWERS

When you add columns of numbers, how do you know that you came up with the right answer? One way to check, or proof, your answer is to add the figures from the bottom to the top. In the problem you just did, start with 0 + 2 in the right (ones) column and work your way up. Then, carry the 2 into the second column and say 2 + 7 + 5 and work your way up again. Your answer should still come out to 270.

ADDING MORE THAN TWO COLUMNS

Now try your hand at adding three columns of numbers.

Problem:

```
    196
    312
    604
    537
    578
    943
  + 725
```

Did you get the correct answer? You'll know for sure if you proofed it. If you haven't, then go back right now and check your work. I'll wait right here.

Solution:

```
   2 3
    196
    312
    604
    537
    578
    943
  + 725
  3,895
```

Did you get it right? Did you get 3,895 for your answer? If you did the problem correctly, then you're ready to move on to subtraction. You may skip the rest of this section, pass GO, collect $200, and go directly to subtracting in the next section.

PROBLEM SET

If you're still a little rusty with your addition, then what you need is some more practice. So I'd like you to do this problem set.

1.		2.		3.	
	209		175		119
	810		316		450
	175		932		561
	461		509		537
	334		140		366
	520		462		914
	312		919		838
	685		627		183
	+ 258		+ 413		+ 925

Solutions

Did you check your answers for each problem? If so, you should have gotten the answers shown below.

1.		2.		3.	
	$\overset{33}{209}$		$\overset{24}{175}$		$\overset{34}{119}$
	810		316		450
	175		932		561
	461		509		537
	334		140		366
	520		462		914
	312		919		838
	685		627		183
	+ 258		+ 413		+ 925
	3,764		4,493		4,893

NEXT STEP

Now that you've mastered addition, you're ready to tackle subtraction. But if you still need a little more practice, then why not redo this lesson? If you've been away from working with numbers for a while, it takes some getting used to.

REVIEW
LESSON | 2

This lesson reviews how to subtract whole numbers. If you missed any of the subtraction questions in the pretest, this lesson will guide you through basic subtraction concepts.

SUBTRACTION

Subtraction is the mathematical opposite of addition. Instead of combining one number with another, we take one away from another. For instance, you might ask someone, "How much is 68 take away 53?" This question is written in the form of problem 1.

SIMPLE SUBTRACTION

First let's start off by working out some basic subtraction problems. These problems are simple because you don't have to borrow or cancel any numbers.

PROBLEM SET

Try these two-column subtraction problems.

1.
$$68 \\ -\,53$$

3.
$$77 \\ -\,36$$

2.
$$94 \\ -\,41$$

4.
$$82 \\ -\,50$$

Solutions

How did you do? You should have gotten the following answers.

1.
$$68 \\ -\,53 \\ \hline 15$$

3.
$$77 \\ -\,36 \\ \hline 41$$

2.
$$94 \\ -\,41 \\ \hline 53$$

4.
$$82 \\ -\,50 \\ \hline 32$$

CHECKING YOUR SUBTRACTION

There's a great way to check or proof your answers. Just add your answer to the number you subtracted and see if they add up to the number you subtracted from.

1.
$$15 \\ +\,53 \\ \hline 68$$

3.
$$41 \\ +\,36 \\ \hline 77$$

2.
$$53 \\ +\,41 \\ \hline 94$$

4.
$$32 \\ +\,50 \\ \hline 82$$

SUBTRACTING WITH BORROWING

Now I'll add a wrinkle. You're going to need to borrow. Are you ready?

PROBLEM SET

Okay? Then find answers to these problems.

5.
$$\begin{array}{r} 54 \\ -\ 49 \\ \hline \end{array}$$

7.
$$\begin{array}{r} 86 \\ -\ 58 \\ \hline \end{array}$$

6.
$$\begin{array}{r} 63 \\ -\ 37 \\ \hline \end{array}$$

8.
$$\begin{array}{r} 97 \\ -\ 49 \\ \hline \end{array}$$

Solutions

5.
$$\begin{array}{r} {}^{4}\!\!{}^{1}54 \\ -\ 49 \\ \hline 5 \end{array}$$

7.
$$\begin{array}{r} {}^{7}\!\!{}^{1}86 \\ -\ 58 \\ \hline 28 \end{array}$$

6.
$$\begin{array}{r} {}^{5}\!\!{}^{1}63 \\ -\ 37 \\ \hline 26 \end{array}$$

8.
$$\begin{array}{r} {}^{8}\!\!{}^{1}97 \\ -\ 49 \\ \hline 48 \end{array}$$

If you got these right, please go directly to the next section, multiplication. And if you didn't? Well, nobody's perfect. But you'll get a lot closer to perfection with a little more practice.

We need to talk about borrowing. In problem 5, we needed to subtract 9 from 4. Well, that's pretty hard to do. So we made the 4 into 14 by borrowing 1 from the 5 of 54. Okay, so 14 − 9 is 5. Since we borrowed 1 from the 5, that 5 is now 4. And 4 − 4 is 0. So 54 − 49 = 5.

Next case. In problem 6, we're subtracting 37 from 63. But 7 is larger than 3, so we borrowed 1 from the 6. That makes the 6 just 5, but it makes 3 into 13. 13 − 7 = 6. And 5 − 3 = 2.

PROBLEM SET

Have you gotten the hang of it? Let's find out. Complete this problem set.

9.
$$\begin{array}{r} 57 \\ -\ 38 \\ \hline \end{array}$$

11.
$$\begin{array}{r} 80 \\ -\ 55 \\ \hline \end{array}$$

10.
$$\begin{array}{r} 72 \\ -\ 29 \\ \hline \end{array}$$

12.
$$\begin{array}{r} 91 \\ -\ 76 \\ \hline \end{array}$$

Solutions

9. $\overset{4\,1}{\cancel{57}}$
-38
$\overline{19}$

11. $\overset{7\,1}{\cancel{80}}$
-55
$\overline{25}$

10. $\overset{6\,1}{\cancel{72}}$
-29
$\overline{43}$

12. $\overset{8\,1}{\cancel{91}}$
-76
$\overline{15}$

SUBTRACTING WITH MORE THAN TWO DIGITS

Now we'll take subtraction just one more step. Do you know how to do the two-step, a dance that's favored in Texas and most other western states? Well now you're going to be doing the subtraction three-step, or at least the three-digit.

PROBLEM SET

Try these three-digit subtraction problems.

13. 532
-149

15. 903
-616

14. 714
-385

16. 840
-162

Solutions

13. $\overset{4\,12\,1}{\cancel{532}}$
-149
$\overline{383}$

15. $\overset{8\,9\,1}{\cancel{903}}$
-616
$\overline{287}$

14. $\overset{6\,10\,1}{\cancel{714}}$
-385
$\overline{329}$

16. $\overset{7\,13\,1}{\cancel{840}}$
-162
$\overline{678}$

Did you get the right answers? Proof them to find out. If you haven't already checked yours, go ahead, and then check your work against mine.

Answer Check

13. $\overset{1\,1}{3}83$
  $+\ 149$
  $\overline{532}$

15. $\overset{1\,1}{2}87$
  $+\ 616$
  $\overline{903}$

14. $\overset{1\,1}{3}29$
  $+\ 385$
  $\overline{714}$

16. $\overset{1\,1}{6}78$
  $+\ 162$
  $\overline{840}$

NEXT STEP

How did you do? If you got everything right, then go directly to Review Lesson 3. But if you feel you need more work subtracting, please redo this lesson.

REVIEW LESSON 3

Multiplication is one of the most important building blocks in mathematics. Without multiplication, you can't do division, elementary algebra, or very much beyond that. The key to multiplication is memorizing the multiplication table found at the end of this lesson.

MULTIPLICATION

Multiplication is addition. For instance, how much is 5 × 4? You know it's 20 because you searched your memory for that multiplication fact. There's nothing wrong with that. As long as you can remember what the answer is from the multiplication table, you're all right.

Another way to calculate 5 × 4 is to add them: 4 + 4 + 4 + 4 + 4 = 20.

We do multiplication instead of addition because it's shorter. Suppose you had to multiply 395 × 438. If you set this up as an addition problem, you'd be working at it for a couple of hours.

SIMPLE MULTIPLICATION

Do you know the multiplication table? You definitely know most of it from 1 × 1 all the way up to 10 × 10. But a lot of people have become so dependent on their calculators that they've forgotten a few of the multi-plication solutions—like 9×6 or 8×7.

Multiplication is basic to understanding mathematics. And to really know how to multiply, you need to know the entire multiplication table by memory.

So I'll tell you what I'm going to do. I'll let you test yourself. First fill in the answers to the multiplication problems in the table that follows. Then check your work against the numbers shown in the completed multiplication table that appears at the end of the lesson. If they match, then you know the entire table. But if you missed a few, then you'll need to practice doing those until you've committed them to memory. Just make up flash cards (with the problem on one side and the answer on the other) for the problems you missed. Once you've done that, they're yours.

Multiplication Table

	1	2	3	4	5	6	7	8	9	10
1	1	2								
2	2	4								
3			9	12						
4										
5										
6										
7										
8										
9										
10										

LONG MULTIPLICATION

We've talked about using calculators before, so remember the deal we made. You should not use calculators for simple arithmetic calculations unless the problems are so repetitive that they become tedious. So I want you to keep working without a calculator.

PROBLEM SET

I'd like you to do this problem set.

1.	46	2.	92	3.	83
	× 37		× 18		× 78

Solutions

1.	46	2.	92	3.	83
	× 37		× 18		× 78
	322		736		664
	138		92		5 81
	1,702		1,656		6,474

You still have to carry numbers in these problems, although they are not shown in the solutions. To show you how it works, I'm going to talk you through the first problem, step-by-step. First we multiply 6 × 7, which gives us 42. We write down the 2 and carry the 4:

$$\begin{array}{r} {}^{4}46 \\ \times\,37 \\ \hline 2 \end{array}$$

Then we multiply 4 × 7, which gives us 28. We add the 4 we carried to the 28 and write down 32:

$$\begin{array}{r} {}^{4}46 \\ \times\,37 \\ \hline 322 \end{array}$$

Next we multiply 6 × 3, giving us 18. We write down the 8 and carry the 1:

$$
\begin{array}{r}
{}^{1}46 \\
\times\, 37 \\
\hline
322 \\
8 \\
\end{array}
$$

Then we multiply 4 × 3, giving us 12. We add the 1 we carried to the 12, and write down 13:

$$
\begin{array}{r}
{}^{1}46 \\
\times\, 37 \\
\hline
322 \\
138 \\
\end{array}
$$

After that we add our columns:

$$
\begin{array}{r}
46 \\
\times\, 37 \\
\hline
322 \\
1\,38 \\
\hline
1,702 \\
\end{array}
$$

Did you get the right answers for the whole problem set? Want to see how to check your answers? Read on for an easy checking system.

HOW TO CHECK YOUR ANSWERS

To prove your multiplication, just reverse the numbers you're multiplying.

1.	2.	3.
$\begin{array}{r} 37 \\ \times\, 46 \\ \hline 222 \\ 1\,48 \\ \hline 1,702 \end{array}$	$\begin{array}{r} 18 \\ \times\, 92 \\ \hline 36 \\ 1\,62 \\ \hline 1,656 \end{array}$	$\begin{array}{r} 78 \\ \times\, 83 \\ \hline 234 \\ 6\,24 \\ \hline 6,474 \end{array}$

If you got these right, then you can skip the section below entitled, "Multiplication: Step-by-Step." But if you're still a little shaky about multiplying, then you should definitely read it.

MULTIPLICATION: STEP-BY-STEP

Long multiplication is just simple multiplication combined with addition. Let's multiply 89 by 57. Here is a step-by-step list describing how to get the answer.

a.   89
         $\times\,57$

b. $7 \times 9 = 63$

c. Write down the 3 and carry the 6.

d.   89 carry 6
         $\times\,57$
            3

e. $7 \times 8 = 56$

f. $56 + 6 = 62$

g. Write down 62.

h.   89
         $\times\,57$
          623

i. $5 \times 9 = 45$

j. Write down the 5 and carry the 4.

k.   89 carry 4
         $\times\,57$
          623
            5

l. $5 \times 8 = 40$

m. $40 + 4 = 44$

n. Write down 44.

o.
```
    89
  × 57
   623
   445
```

p. Add the two numbers you got.
```
    89
  × 57
   623
  4 45
  5,073
```

PROBLEM SET

Let's try something a little harder. Do these multiplication problems.

4.
```
    537
  × 219
```

5.
```
    954
  × 628
```

6.
```
    791
  × 524
```

Solutions

4.
```
    537
  × 219
  4 833
  5 37
  107 4
  117,603
```

5.
```
    954
  × 628
  7 632
  19 08
  572 4
  599,112
```

6.
```
    791
  × 524
  3 164
  15 82
  395 5
  414,484
```

Will you get the same answer multiplying 111 × 532 as you will if you multiply 532 × 111? Let's find out. Please work out both problems:

```
    111              532
  × 532            × 111
```

Solutions

```
    111              532
  × 532            × 111
    222              532
    333              532
    555              532
 59,052           59,052
```

We get the same answer both ways. So you always have a choice when you multiply. In this case, is it easier to multiply 111 × 532 or 532 × 111?

Obviously it's much easier to multiply 532 × 111, because you don't really do any multiplying. All you do is write 532 three times, and then add.

OK, which is easier, multiplying 749 × 222 or 222 × 749? Please work it out both ways:

$$\begin{array}{r} 749 \\ \times\, 222 \\ \hline \end{array} \qquad\qquad \begin{array}{r} 222 \\ \times\, 749 \\ \hline \end{array}$$

Solutions

$$\begin{array}{r} 749 \\ \times\, 222 \\ \hline 1498 \\ 1498 \\ \underline{1498} \\ 166{,}278 \end{array} \qquad\qquad \begin{array}{r} 222 \\ \times\, 749 \\ \hline 1998 \\ 888 \\ \underline{1554} \\ 166{,}278 \end{array}$$

You can see that multiplying 749 × 222 is quite a bit easier than multiplying 222 × 749. As you do more and more problems, you'll recognize shortcuts like this one.

Next Step

Okay, no more Mr. Nice Guy. Because mastering multiplication is so important, I must insist that you really have this down before you go on to division. After all, if you can't multiply, then you can't divide. It's as simple as that. So if you got any of the problems in this section wrong, go back and work through them again. And memorize your multiplication table!

Completed Multiplication Table

	1	2	3	4	5	6	7	8	9	10
1	1	2	3	4	5	6	7	8	9	10
2	2	4	6	8	10	12	14	16	18	20
3	3	6	9	12	15	18	21	24	27	30
4	4	8	12	16	20	24	28	32	36	40
5	5	10	15	20	25	30	35	40	45	50
6	6	12	18	24	30	36	42	48	54	60
7	7	14	21	28	35	42	49	56	63	70
8	8	16	24	32	40	48	56	64	72	80
9	9	18	27	36	45	54	63	72	81	90
10	10	20	30	40	50	60	70	80	90	100

REVIEW LESSON | 4

This lesson will help you master both short and long division. It will also show you how to check, or proof, your answers, so you can know for certain that the answer you came up with is indeed correct.

DIVISION

As you'll see, division is the opposite of multiplication. So you really must know the multiplication table from the previous lesson to do these division problems. In this lesson, you'll learn the difference between short and long division and how to use trial and error to get to the solution.

SHORT DIVISION
We'll start you off with a set of short division problems.

PROBLEM SET
Work out the answers to the problems on the next page.

1. $5 \overline{)\,140}$ 2. $9 \overline{)\,189}$ 3. $7 \overline{)\,2{,}114}$

Solutions

1.
$$5 \overline{)\,1\overset{4}{4}0}^{\,28}$$

2.
$$9 \overline{)\,189}^{\,21}$$

3.
$$7 \overline{)\,2{,}114}^{\,302}$$

Let's take a closer look at problem 3. We divide 7 into 21 to get the 3:

$$7 \overline{)\,2{,}114}^{\,3}$$

Then we try to divide 7 into 1. Since 7 is larger than 1, it doesn't fit. So we write 0 over the 1:

$$7 \overline{)\,2{,}114}^{\,30}$$

And then we ask how many times 7 goes into 14. The answer is 2:

$$7 \overline{)\,2{,}114}^{\,302}$$

HOW TO CHECK YOUR ANSWERS

The answers to each of these problems can be checked, or proven. I'll do the first proof below.

1.
$$\begin{array}{r} 28 \\ \times\,5 \\ \hline 140 \end{array}$$

Now you do the next proofs.

Did your answers check out? Here are my proofs.

2.
$$
\begin{array}{r}
21 \\
\times\,9 \\
\hline
189
\end{array}
$$

3.
$$
\begin{array}{r}
302 \\
\times\,7 \\
\hline
2{,}114
\end{array}
$$

Each of these came out even. But sometimes there's a remainder. You'll find that that's the case in the next problem set. When you learn about decimals in Section III, you'll find out you can keep dividing until it comes out even, or you can round off the answer.

PROBLEM SET

Now try these division problems that don't come out even. They all have remainders.

4. $9\,\overline{)\,413}$ **5.** $8\,\overline{)\,321}$ **6.** $6\,\overline{)\,501}$

Solutions

4. $9\,\overline{)\,41\overset{5}{3}}$ 45 R8 **5.** $8\,\overline{)\,321}$ 40 R1 **6.** $6\,\overline{)\,50\overset{2}{1}}$ 83 R3

LONG DIVISION

Long division is carried out in two steps:
- Trial and error
- Multiplication

The process of long division is identical to short division, but it involves a lot more calculation. That's why it's so important to have memorized the multiplication table.

Let's work out the next problem together.

Problem: $37\,\overline{)\,596}$

Solution: How many times does 37 go into 59? Just once. So we put a 1 directly over the 9 and write in 37 directly below 59.

$$
\begin{array}{r}
1 \\
37\,\overline{)\,596} \\
-\,37 \\
\hline
22
\end{array}
$$

Then we subtract 37 from 59, leaving us with 22. Next, we bring down the 6, giving us 226. How many times does 37 go into 226? We need to do this by trial and error. We finally come up with 6, since $6 \times 37 = 222$.

$$
\begin{array}{r}
16 \\
37 \overline{)596} \\
-37X \\
\hline
226 \\
-222 \\
\end{array}
$$

When we subtract 222 from 226, we are left with 4, which is our remainder.

$$
\begin{array}{r}
16 \\
37 \overline{)596} \\
-37X \\
\hline
226 \\
-222 \\
\hline
4 \\
\end{array}
$$

The proper notation for the answer is 16 R4. Can you check this answer? Yes! Just multiply 16×37 and add 4. Go ahead and do it now. Did you get 596? Good. Then you proved your answer, 16 R4, is correct.

Here's another problem. Find the answer and then check it.

Problem: $43 \overline{)985}$

Solution:

$$
\begin{array}{r}
22 \text{ R39} \\
43 \overline{)985} \\
-86X \\
\hline
125 \\
-86 \\
\hline
39 \\
\end{array}
$$

Proof:

$$
\begin{array}{r}
43 \\
\times\,22 \\
\hline
86 \\
\underline{86} \\
946 \\
\underline{+\,39} \\
985
\end{array}
$$

PROBLEM SET

Here's a problem set for you to work on.

7.　　$86\,\overline{)\,4{,}135}$　　　　　**9.**　$116\,\overline{)\,7{,}048}$

8.　　$93\,\overline{)\,2{,}740}$　　　　　**10.** $235\,\overline{)\,91{,}538}$

Solutions

7.
$$
\begin{array}{r}
48\ \ \text{R7} \\
86\,\overline{)\,4{,}135} \\
-\ \underline{3\,44\text{X}} \\
695 \\
-\ \underline{688} \\
7
\end{array}
$$

9.
$$
\begin{array}{r}
60\ \ \text{R88} \\
116\,\overline{)\,7{,}048} \\
-\ \underline{6\,96\text{X}} \\
88
\end{array}
$$

8.
$$
\begin{array}{r}
29\ \ \text{R43} \\
93\,\overline{)\,2{,}740} \\
-\ \underline{1\,86\text{X}} \\
880 \\
-\ \underline{837} \\
43
\end{array}
$$

10.
$$
\begin{array}{r}
389\ \ \text{R123} \\
235\,\overline{)\,91{,}538} \\
-\ \underline{70\,5\text{XX}} \\
21\,03 \\
-\ \underline{18\,80} \\
2\,238 \\
-\ \underline{2\,115} \\
123
\end{array}
$$

NEXT STEP

Have you been tempted to reach for your calculator to do some of the problems in this section? Remember that the less you rely on your calculator, and the more you rely on your own mathematical ability, the better off you'll be. The more you rely on your ability, the more your ability will be developed.

If you've mastered addition, subtraction, multiplication, and division of whole numbers, you're ready to tackle fractions.

SECTION | II

FRACTIONS

How many times a day do you hear ads on television—especially on the home shopping channels—offering you some pretty amazing products at just a **fraction** of what you would have to pay for them in a store? Of course you need to ask just what kind of fraction they're talking about. Is it $\frac{1}{2}$, $\frac{1}{3}$, $\frac{1}{4}$, or $\frac{9}{10}$?

We'll start with the fraction $\frac{1}{2}$. The top number is called the *numerator* and the bottom number is called the *denominator*. So in the fraction $\frac{1}{2}$, the numerator is 1 and the denominator is 2. In the fraction $\frac{2}{3}$, the numerator is 2 and the denominator is 3.

In a *proper fraction* the denominator is always greater than the numerator. We already saw that $\frac{1}{2}$ and $\frac{1}{3}$ are proper fractions. How about $\frac{4}{5}$, $\frac{3}{8}$, and $\frac{19}{20}$? These too, are proper fractions.

What do you think *improper fractions* look like? They look like these fractions: $\frac{2}{1}$, $\frac{17}{14}$, and $\frac{7}{5}$. So if the numerator is greater than the denominator, then it's an improper fraction.

What if the numerator and the denominator are equal (making the fraction equal to 1), as is the case with these fractions: $\frac{2}{2}$, $\frac{9}{9}$, $\frac{20}{20}$? Are these proper or improper fractions? A while back someone decided that when the numerator and denominator are equal, we must call that an improper fraction. That's the rule, but it's not really all that important.

What *is* important is to recognize the relationship between the numerator and the denominator. Let's take the improper fraction $\frac{4}{2}$. What are you supposed to do with it? Should we just leave it sitting there? Or maybe do a little division? Okay, you do a little division. Now what do you divide into what? You divide the 2 into the 4, which gives you 2.

So the relationship of the numerator to the denominator of a fraction is that you're supposed to divide the denominator (or bottom) of the fraction into the numerator (or top).

There's one more term I'd like to introduce, and then we can stop talking about fractions and start using them. The term is *mixed number*, which consists of a whole number and a proper fraction. Examples would include numbers like $3\frac{3}{4}$, $1\frac{5}{8}$, and $4\frac{2}{3}$.

Do you really have to know all these terms? Not necessarily. Just remember *numerator* and *denominator*. If you can also remember *proper fraction, improper fraction,* and *mixed number,* then you will have enriched your vocabulary, but you'll still have to get out of bed every morning, and you probably won't notice any major changes in the quality of your life.

When you have completed this section, you will be able to convert improper fractions into mixed numbers and convert mixed numbers into improper fractions. You'll also be able to add, subtract, multiply, and divide proper fractions, improper fractions, and mixed numbers.

LESSON | 1

In this lesson, you'll
learn the basic fraction
conversion procedures.
These procedures will be
used when you move on to
the more complicated
fraction problems,
so be sure to read this
lesson carefully.

FRACTION
CONVERSIONS

By convention, answers to fraction problems are expressed in terms of mixed numbers, rather than in terms of improper fractions. But when you add, subtract, multiply, and divide mixed numbers—which you'll be doing later in this section—you'll find it a lot easier to work with improper fractions. So you need to be able to convert improper fractions into mixed numbers and mixed numbers into improper fractions.

CONVERTING IMPROPER FRACTIONS INTO MIXED NUMBERS

To convert an improper fraction into a mixed number, you divide the denominator (bottom number) into the numerator (top number). Any remainder becomes the numerator of the fraction part of the mixed number.

Problem: Can you convert $\frac{5}{2}$ into a mixed number?

Solution: $\frac{5}{2} = 2\frac{1}{2}$

Here is another one for you to try.

Problem: Convert $\frac{7}{3}$ into a mixed number.

Solution: $\frac{7}{3} = 2\frac{1}{3}$

Try one more.

Problem: Convert $\frac{24}{9}$ into a mixed number.

Solution: $\frac{24}{9} = 2\frac{6}{9} = 2\frac{2}{3}$

You generally need to reduce your fractions to the lowest possible terms. In other words, get the denominator as low as possible. You do this by dividing both the numerator and the denominator by the same number. In this case, I divided both 6 and 9 by 3 to change $\frac{6}{9}$ into $\frac{2}{3}$.

PROBLEM SET
Convert each of these improper fractions into mixed numbers.

1. $\frac{9}{2} =$

2. $\frac{15}{4} =$

3. $\frac{12}{7} =$

4. $\frac{26}{4} =$

5. $\frac{34}{6} =$

6. $\frac{19}{5} =$

Solutions

1. $\frac{9}{2} = 4\frac{1}{2}$

2. $\frac{15}{4} = 3\frac{3}{4}$

3. $\frac{12}{7} = 1\frac{5}{7}$

4. $\frac{26}{4} = 6\frac{2}{4} = 6\frac{1}{2}$

5. $\frac{34}{6} = 5\frac{4}{6} = 5\frac{2}{3}$

6. $\frac{19}{5} = 3\frac{4}{5}$

CONVERTING MIXED NUMBERS INTO IMPROPER FRACTIONS

We've converted improper fractions into mixed numbers, so for our next trick, we're going to convert mixed numbers into improper fractions. You need to follow a two-step process:

1. Multiply the whole number by the denominator of the fraction.
2. Add that number (or product) to the numerator of the fraction.

PROBLEM SET

Convert each of these mixed numbers into improper fractions.

7. $3\frac{4}{5} =$

8. $5\frac{4}{7} =$

9. $1\frac{9}{10} =$

10. $1\frac{2}{3} =$

11. $5\frac{1}{6} =$

12. $4\frac{5}{7} =$

13. $10\frac{2}{3} =$

14. $12\frac{9}{10} =$

15. $15\frac{3}{4} =$

Solutions

7. $3\frac{4}{5} = \frac{19}{5}$ $(3 \times 5 = 15; 15 + 4 = 19)$

8. $5\frac{4}{7} = \frac{39}{7}$ $(5 \times 7 = 35; 35 + 4 = 39)$

9. $1\frac{9}{10} = \frac{19}{10}$ $(1 \times 10 = 10; 10 + 9 = 19)$

10. $1\frac{2}{3} = \frac{5}{3}$

11. $5\frac{1}{6} = \frac{31}{6}$

12. $4\frac{5}{7} = \frac{33}{7}$

13. $10\frac{2}{3} = \frac{32}{3}$

14. $12\frac{9}{10} = \frac{129}{10}$

15. $15\frac{3}{4} = \frac{63}{4}$

NEXT STEP

Converting improper fractions into mixed numbers and mixed numbers into improper fractions are skills you'll be using for the rest of this section. When you're confident that you've mastered these skills, go on to the next lesson. But any time you're not sure you've really got something down, just go back over it. Remember that you're covering a whole lot of math in just 30 days.

LESSON | 2

First, you're going to be adding fractions with the same denominators, and then you'll move on to fractions with different denominators. When you have completed this lesson, you'll be able to add *any* fractions and find the right answer.

ADDING FRACTIONS

Do you have any loose change? I'd like to borrow a quarter. Thanks. Do you happen to have another quarter I could borrow? Don't worry, it's just a loan. And while you're at it, let me borrow still another quarter. All right, then, how many quarters do I owe you?

If I borrowed one quarter from you, then another quarter, and then still another quarter, I borrowed three quarters from you. In other words I borrowed $\frac{1}{4} + \frac{1}{4} + \frac{1}{4}$, or a total of $\frac{3}{4}$.

Now before I forget, let me return those three quarters.

ADDING FRACTIONS WITH COMMON DENOMINATORS

Here's another question: How much is $\frac{1}{10} + \frac{1}{10} + \frac{1}{10}$? It's $\frac{3}{10}$. And how much is $\frac{2}{9} + \frac{2}{9} + \frac{2}{9} + \frac{2}{9}$? Go ahead and add them up. It's $\frac{8}{9}$. When you add fractions with the same denominator, all you have to do is add the numerators.

How much is $\frac{1}{6} + \frac{1}{6} + \frac{1}{6}$? It's $\frac{3}{6}$. But we can reduce that to $\frac{1}{2}$. What did we really do just then? We divided the numerator (3) by 3 and we divided the denominator (6) by 3. There's a law of arithmetic that says when you divide the top of a fraction by any number, you must also divide the bottom of that fraction by the same number.

Now add together $\frac{1}{2} + \frac{1}{2} + \frac{1}{2} + \frac{1}{2}$. What did you come up with? Was it 2? All **right**! You did this: $\frac{1}{2} + \frac{1}{2} + \frac{1}{2} + \frac{1}{2} = \frac{4}{2} = 2$.

PROBLEM SET

Here's a set of problems for you to solve.

1. $\frac{1}{5} + \frac{2}{5} + \frac{2}{5} =$

2. $\frac{1}{9} + \frac{2}{9} + \frac{3}{9} =$

3. $\frac{1}{8} + \frac{1}{8} + \frac{2}{8} + \frac{2}{8} =$

4. $\frac{2}{12} + \frac{3}{12} + \frac{1}{12} + \frac{2}{12} =$

5. $\frac{1}{20} + \frac{3}{20} + \frac{2}{20} + \frac{4}{20} + \frac{1}{20} =$

6. $\frac{2}{50} + \frac{3}{50} + \frac{7}{50} + \frac{4}{50} + \frac{8}{50} =$

Solutions

1. $\frac{1}{5} + \frac{2}{5} + \frac{2}{5} = 1$

2. $\frac{1}{9} + \frac{2}{9} + \frac{3}{9} = \frac{6}{9} = \frac{2}{3}$

3. $\frac{1}{8} + \frac{1}{8} + \frac{2}{8} + \frac{2}{8} = \frac{6}{8} = \frac{3}{4}$

4. $\frac{2}{12} + \frac{3}{12} + \frac{1}{12} + \frac{2}{12} = \frac{8}{12} = \frac{2}{3}$

5. $\frac{1}{20} + \frac{3}{20} + \frac{2}{20} + \frac{4}{20} + \frac{1}{20} = \frac{11}{20}$

6. $\frac{2}{50} + \frac{3}{50} + \frac{7}{50} + \frac{4}{50} + \frac{8}{50} = \frac{24}{50} = \frac{12}{25}$

Did you reduce all your fractions to their lowest possible terms? If you left problem 1 at $\frac{5}{5}$, is it wrong? No, but by convention we always reduce our fractions as much as possible. Indeed, there are mathematicians who can't go to sleep at night unless they're sure that every fraction has been reduced to its lowest possible terms. Now I'm sure that you wouldn't want to keep these poor people up all night, so always reduce your fractions.

ADDING FRACTIONS WITH UNLIKE DENOMINATORS

So far we've been adding fractions with common denominators—halves, quarters, sixths, tenths, and so forth. Now we'll be adding fractions that don't have common denominators.

Have you ever heard the expression, "That's like adding apples and oranges?" You can add apples and apples—3 apples plus 2 apples equal 5 apples. And you can add oranges—4 oranges plus 3 oranges equal 7 oranges. But you can't add apples *and* oranges.

Can you add $\frac{1}{2}$ and $\frac{1}{3}$? Believe it or not, you can. The problem is that they don't have a common denominator. In the last problem set the fractions in each problem had a common denominator. In problem 1 you added $\frac{1}{5} + \frac{2}{5} + \frac{2}{5}$. In problem 2 you added $\frac{1}{9} + \frac{2}{9} + \frac{3}{9}$. And in problem 3 you added $\frac{1}{8} + \frac{1}{8} + \frac{2}{8} + \frac{2}{8}$.

What we need to do to add $\frac{1}{2}$ and $\frac{1}{3}$ is to give them a common denominator. Do you have any ideas? Think about it for a while.

All right, time's up! Did you think of converting $\frac{1}{2}$ into $\frac{3}{6}$? And $\frac{1}{3}$ into $\frac{2}{6}$? Here's how you could do it: $\frac{1 \times 3}{2 \times 3} + \frac{1 \times 2}{3 \times 2} = \frac{3}{6} + \frac{2}{6} = \frac{5}{6}$.

Remember that old arithmetic law: What you do to the bottom of a fraction (the denominator), you must also do to the top (the numerator).

Once the fractions have a common denominator, you can add them: $\frac{3}{6} + \frac{2}{6} = \frac{5}{6}$.

Try your hand at adding the following two fractions.

Problem: $\frac{1}{3} + \frac{1}{5} =$

Solution: $\frac{1}{3} + \frac{1}{5} = \frac{1 \times 5}{3 \times 5} + \frac{1 \times 3}{5 \times 3} = \frac{5}{15} + \frac{3}{15} = \frac{8}{15}$

Problem Set

Here's a problem set to work out.

7. $\frac{1}{4} + \frac{1}{3} =$

8. $\frac{1}{3} + \frac{1}{6} =$

9. $\frac{1}{6} + \frac{1}{4} =$

10. $\frac{1}{4} + \frac{2}{5} =$

11. $\frac{3}{10} + \frac{2}{5} =$

12. $\frac{3}{8} + \frac{5}{12} =$

Solutions

7. $\frac{1}{4} + \frac{1}{3} = \frac{1 \times 3}{4 \times 3} + \frac{1 \times 4}{3 \times 4} = \frac{3}{12} + \frac{4}{12} = \frac{7}{12}$

8. $\frac{1}{3} + \frac{1}{6} = \frac{1 \times 2}{3 \times 2} + \frac{1}{6} = \frac{2}{6} + \frac{1}{6} = \frac{3}{6} = \frac{1}{2}$

9. $\frac{1}{6} + \frac{1}{4} = \frac{1 \times 2}{6 \times 2} + \frac{1 \times 3}{4 \times 3} = \frac{2}{12} + \frac{3}{12} = \frac{5}{12}$

10. $\frac{1}{4} + \frac{2}{5} = \frac{1 \times 5}{4 \times 5} + \frac{2 \times 4}{5 \times 4} = \frac{5}{20} + \frac{8}{20} = \frac{13}{20}$

11. $\frac{3}{10} + \frac{2}{5} = \frac{3}{10} + \frac{2 \times 2}{5 \times 2} = \frac{3}{10} + \frac{4}{10} = \frac{7}{10}$

12. $\frac{3}{8} + \frac{5}{12} = \frac{3 \times 3}{8 \times 3} + \frac{5 \times 2}{12 \times 2} = \frac{9}{24} + \frac{10}{24} = \frac{19}{24}$

In problem 9, if you did it the way I did it below, it's okay. By not finding the lowest common denominator, you needed to do an extra step—which doesn't matter if you ended up with the right answer.

$$\frac{1}{6} + \frac{1}{4} = \frac{1 \times 4}{6 \times 4} + \frac{1 \times 6}{4 \times 6} = \frac{4}{24} + \frac{6}{24} = \frac{10}{24} = \frac{5}{12}$$

ADDING SEVERAL FRACTIONS TOGETHER

So far we've been adding two fractions. Can we add three or four fractions the same way? We definitely can—and will. See what you can do with this one:

Problem: $\frac{1}{4} + \frac{1}{5} + \frac{1}{20} =$

Solution: $\frac{1}{4} + \frac{1}{5} + \frac{1}{20} = \frac{1 \times 5}{4 \times 5} + \frac{1 \times 4}{5 \times 4} + \frac{1}{20} = \frac{5}{20} + \frac{4}{20} + \frac{1}{20} = \frac{10}{20} = \frac{1}{2}$

Here's one more.

Problem: $\frac{1}{8} + \frac{2}{5} + \frac{1}{4} + \frac{3}{20} =$

Solution: $\frac{1}{8} + \frac{2}{5} + \frac{1}{4} + \frac{3}{20} = \frac{1 \times 5}{8 \times 5} + \frac{2 \times 8}{5 \times 8} + \frac{1 \times 10}{4 \times 10} + \frac{3 \times 2}{20 \times 2}$

$= \frac{5}{40} + \frac{16}{40} + \frac{10}{40} + \frac{6}{40} = \frac{37}{40}$

PROBLEM SET

Try a problem set with more than two fractions.

13. $\frac{1}{10} + \frac{1}{3} + \frac{1}{5} + \frac{1}{6} =$

14. $\frac{1}{12} + \frac{1}{4} + \frac{1}{3} =$

15. $\frac{3}{20} + \frac{1}{4} + \frac{1}{5} =$

16. $\frac{1}{5} + \frac{1}{4} + \frac{1}{8} + \frac{7}{20} =$

17. $\frac{2}{15} + \frac{1}{5} + \frac{1}{6} + \frac{3}{10} =$

18. $\frac{1}{3} + \frac{1}{12} + \frac{1}{4} + \frac{1}{6} =$

Solutions

13. $\frac{1}{10} + \frac{1}{3} + \frac{1}{5} + \frac{1}{6} = \frac{1 \times 3}{10 \times 3} + \frac{1 \times 10}{3 \times 10} + \frac{1 \times 6}{5 \times 6} + \frac{1 \times 5}{6 \times 5} = \frac{3}{30} + \frac{10}{30} + \frac{6}{30} + \frac{5}{30}$

$= \frac{24}{30} = \frac{4}{5}$

14. $\frac{1}{12} + \frac{1}{4} + \frac{1}{3} = \frac{1}{12} + \frac{1 \times 3}{4 \times 3} + \frac{1 \times 4}{3 \times 4} = \frac{1}{12} + \frac{3}{12} + \frac{4}{12} = \frac{8}{12} = \frac{2}{3}$

15. $\frac{3}{20} + \frac{1}{4} + \frac{1}{5} = \frac{3}{20} + \frac{1 \times 5}{4 \times 5} + \frac{1 \times 4}{5 \times 4} = \frac{3}{20} + \frac{5}{20} + \frac{4}{20} = \frac{12}{20} = \frac{3}{5}$

16. $\frac{1}{5} + \frac{1}{4} + \frac{1}{8} + \frac{7}{20} = \frac{1 \times 8}{5 \times 8} + \frac{1 \times 10}{4 \times 10} + \frac{1 \times 5}{8 \times 5} + \frac{7 \times 2}{20 \times 2}$

$= \frac{8}{40} + \frac{10}{40} + \frac{5}{40} + \frac{14}{40} = \frac{37}{40}$

17. $\frac{2}{15} + \frac{1}{5} + \frac{1}{6} + \frac{3}{10} = \frac{2 \times 2}{15 \times 2} + \frac{1 \times 6}{5 \times 6} + \frac{1 \times 5}{6 \times 5} + \frac{3 \times 3}{10 \times 3}$

$= \frac{4}{30} + \frac{6}{30} + \frac{5}{30} + \frac{9}{30} = \frac{24}{30} = \frac{4}{5}$

18. $\frac{1}{3} + \frac{1}{12} + \frac{1}{4} + \frac{1}{6} = \frac{1 \times 4}{3 \times 4} + \frac{1}{12} + \frac{1 \times 3}{4 \times 3} + \frac{1 \times 2}{6 \times 2}$

$= \frac{4}{12} + \frac{1}{12} + \frac{3}{12} + \frac{2}{12} = \frac{10}{12} = \frac{5}{6}$

Do you really need to write out all these steps? Let's take another look at problem 18. Maybe we can skip that second step, so our solution would look like this:

$\frac{1}{3} + \frac{1}{12} + \frac{1}{4} + \frac{1}{6} = \frac{4}{12} + \frac{1}{12} + \frac{3}{12} + \frac{2}{12} = \frac{10}{12} = \frac{5}{6}$

And our solution to problem 17 would look like this:

$\frac{2}{15} + \frac{1}{5} + \frac{1}{6} + \frac{3}{10} = \frac{4}{30} + \frac{6}{30} + \frac{5}{30} + \frac{9}{30} = \frac{24}{30} = \frac{4}{5}$

NEXT STEP

Wasn't adding fractions a lot of fun? You'll find that subtracting fractions is an equal amount of fun, and just as easy.

LESSON | 3

This lesson first shows you how to subtract fractions with the same denominator and then moves on to show you how to subtract fractions with different denominators.

SUBTRACTING FRACTIONS

What the Lord giveth, the Lord taketh away. And what holds true in the Bible holds true in mathematics as well. You'll find there's virtually no difference between addition and subtraction except for a change of sign.

SUBTRACTING FRACTIONS WITH COMMON DENOMINATORS

I'm going to start you off with an easy one.

Problem: $\frac{6}{7} - \frac{2}{7} =$

Solution: $\frac{6}{7} - \frac{2}{7} = \frac{4}{7}$

PROBLEM SET

Here is a problem set for you to complete.

1. $\frac{3}{5} - \frac{2}{5} =$

2. $\frac{8}{9} - \frac{1}{9} =$

3. $\frac{17}{20} - \frac{9}{20} =$

4. $\frac{5}{12} - \frac{2}{12} =$

5. $\frac{17}{19} - \frac{4}{19} =$

6. $\frac{9}{10} - \frac{4}{10} =$

Solutions

1. $\frac{3}{5} - \frac{2}{5} = \frac{1}{5}$

2. $\frac{8}{9} - \frac{1}{9} = \frac{7}{9}$

3. $\frac{17}{20} - \frac{9}{20} = \frac{8}{20} = \frac{2}{5}$

4. $\frac{5}{12} - \frac{2}{12} = \frac{3}{12} = \frac{1}{4}$

5. $\frac{17}{19} - \frac{4}{19} = \frac{13}{19}$

6. $\frac{9}{10} - \frac{4}{10} = \frac{5}{10} = \frac{1}{2}$

SUBTRACTING FRACTIONS WITH UNLIKE DENOMINATORS

Let's step back for a minute and take stock. When we added fractions with different denominators, we found their common denominators and added. We do the same thing, then, when we do subtraction with fractions having different denominators.

Problem: How much is $\frac{1}{3} - \frac{1}{4}$?

Solution: $\frac{1}{3} - \frac{1}{4} = \frac{1 \times 4}{3 \times 4} - \frac{1 \times 3}{4 \times 3} = \frac{4}{12} - \frac{3}{12} = \frac{1}{12}$

Here's another one.

Problem: $\frac{1}{5} - \frac{1}{10} =$

Solution: $\frac{1}{5} - \frac{1}{10} = \frac{1 \times 2}{5 \times 2} - \frac{1}{10} = \frac{2}{10} - \frac{1}{10} = \frac{1}{10}$

PROBLEM SET

Are you ready for another problem set? All right, then, let's see what you can do with these problems.

7. $\frac{1}{4} - \frac{1}{8} =$

8. $\frac{1}{3} - \frac{1}{9} =$

9. $\frac{1}{4} - \frac{1}{5} =$

10. $\frac{1}{6} - \frac{1}{16} =$

11. $\frac{1}{2} - \frac{1}{7} =$

12. $\frac{1}{6} - \frac{1}{8} =$

Solutions

7. $\frac{1}{4} - \frac{1}{8} = \frac{1 \times 2}{4 \times 2} - \frac{1}{8} = \frac{2}{8} - \frac{1}{8} = \frac{1}{8}$

8. $\frac{1}{3} - \frac{1}{9} = \frac{1 \times 3}{3 \times 3} - \frac{1}{9} = \frac{3}{9} - \frac{1}{9} = \frac{2}{9}$

9. $\frac{1}{4} - \frac{1}{5} = \frac{1 \times 5}{4 \times 5} - \frac{1 \times 4}{5 \times 4} = \frac{5}{20} - \frac{4}{20} = \frac{1}{20}$

10. $\frac{1}{6} - \frac{1}{16} = \frac{1 \times 8}{6 \times 8} - \frac{1 \times 3}{16 \times 3} = \frac{8}{48} - \frac{3}{48} = \frac{5}{48}$

11. $\frac{1}{2} - \frac{1}{7} = \frac{1 \times 7}{2 \times 7} - \frac{1 \times 2}{7 \times 2} = \frac{7}{14} - \frac{2}{14} = \frac{5}{14}$

12. $\frac{1}{6} - \frac{1}{8} = \frac{1 \times 4}{6 \times 4} - \frac{1 \times 3}{8 \times 3} = \frac{4}{24} - \frac{3}{24} = \frac{1}{24}$

Remember the shortcut we took a few pages ago when we added fractions? We can apply that same shortcut when we subtract fractions. Let's use it for problem 12:

$$\frac{1}{6} - \frac{1}{8} = = \frac{4}{24} - \frac{3}{24} = \frac{1}{24}$$

MORE SUBTRACTION PRACTICE

Now let's try some more complicated subtraction problems. Since you can now subtract fractions which have the number 1 as the numerator, you're ready to try fractions that don't have 1 as the numerator. You'll see that it's the same procedure, but it just takes a few more steps.

Problem: Subtract $\frac{1}{6}$ from $\frac{2}{5}$.

Solution: $\frac{2}{5} - \frac{1}{6} = \frac{2 \times 6}{5 \times 6} - \frac{1 \times 5}{6 \times 5} = \frac{12}{30} - \frac{5}{30} = \frac{7}{30}$

Here's one more.

Problem: $\frac{7}{8} - \frac{3}{5} =$

Solution: $\frac{7}{8} - \frac{3}{5} = \frac{7 \times 5}{8 \times 5} - \frac{3 \times 8}{5 \times 8} = \frac{35}{40} - \frac{24}{40} = \frac{11}{40}$

PROBLEM SET

Here are some more complicated subtraction problems for you to complete.

13. $\frac{3}{5} - \frac{2}{7} =$

14. $\frac{7}{9} - \frac{3}{4} =$

15. $\frac{1}{2} - \frac{3}{8} =$

16. $\frac{5}{7} - \frac{1}{3} =$

17. $\frac{3}{4} - \frac{2}{5} =$

18. $\frac{17}{20} - \frac{5}{6} =$

Solutions

13. $\frac{3}{5} - \frac{2}{7} = \frac{3 \times 7}{5 \times 7} - \frac{2 \times 5}{7 \times 5} = \frac{21}{35} - \frac{10}{35} = \frac{11}{35}$

14. $\frac{7}{9} - \frac{3}{4} = \frac{7 \times 4}{9 \times 4} - \frac{3 \times 9}{4 \times 9} = \frac{28}{36} - \frac{27}{36} = \frac{1}{36}$

15. $\frac{1}{2} - \frac{3}{8} = \frac{1 \times 4}{2 \times 4} - \frac{3}{8} = \frac{4}{8} - \frac{3}{8} = \frac{1}{8}$

16. $\frac{5}{7} - \frac{1}{3} = \frac{5 \times 3}{7 \times 3} - \frac{1 \times 7}{3 \times 7} = \frac{15}{21} - \frac{7}{21} = \frac{8}{21}$

17. $\frac{3}{4} - \frac{2}{5} = \frac{3 \times 5}{4 \times 5} - \frac{2 \times 4}{5 \times 4} = \frac{15}{20} - \frac{8}{20} = \frac{7}{20}$

18. $\frac{17}{20} - \frac{5}{6} = \frac{17 \times 3}{20 \times 3} - \frac{5 \times 10}{6 \times 10} = \frac{51}{60} - \frac{50}{60} = \frac{1}{60}$

NEXT STEP

Believe it or not, you've done all the heavy lifting in this section. As long as you're sure you know how to add and subtract fractions, multiplying and dividing fractions should be a walk in the park.

LESSON | 4

If you know how to
multiply, then you know
how to multiply fractions.
Basically, all you do is
multiply the numerators by
the numerators and the
denominators by the
denominators.

MULTIPLYING FRACTIONS

You'll find that multiplying fractions is different from adding and subtracting them because you don't need to find a common denominator before you do the math operation. Actually, this makes multiplying fractions easier than adding or subtracting them.

EASY MULTIPLICATIONS

How much is one-eighth of a quarter? This is a straightforward multiplication problem. So let's set it up.

Problem: Write down one-eighth as a fraction. Then write down one-quarter.

Solution: Your fractions should look like this: $\frac{1}{8}$, $\frac{1}{4}$.

Problem: The final step is to multiply them. Give it a try and see what you come up with.

Solution: $\frac{1}{8} \times \frac{1}{4} = \frac{1}{32}$

A nice thing about multiplying fractions is that it's not necessary to figure out a common denominator, because you'll find it automatically. But is the 32 in the previous problem the *lowest* common denominator? It is, in this case. Later in this lesson, you'll find that when you multiply fractions, you can often reduce your result to a lower denominator. But for now, let's try another problem that doesn't require you to reduce.

Problem: How much is one-third of one-eighth?

Solution: $\frac{1}{3} \times \frac{1}{8} = \frac{1}{24}$

You can see by the way I'm asking the question that *of* means multiply, or times. The question would be the same if I said, "How much is one-third times one-eighth?"

PROBLEM SET

Try these problems, keeping in mind what the word *of* means in the following questions.

1. Find one-fifth of a quarter.

2. Find one-half of one-third.

3. Find one-eighth of one-half.

4. Find one-quarter of one-sixth.

5. Find one-sixth of one-third.

6. Find one-fifth of one-fifth.

Solutions

1. $\frac{1}{5} \times \frac{1}{4} = \frac{1}{20}$

2. $\frac{1}{2} \times \frac{1}{3} = \frac{1}{6}$

3. $\frac{1}{8} \times \frac{1}{2} = \frac{1}{16}$

4. $\frac{1}{4} \times \frac{1}{6} = \frac{1}{24}$

5. $\frac{1}{6} \times \frac{1}{3} = \frac{1}{18}$

6. $\frac{1}{5} \times \frac{1}{5} = \frac{1}{25}$

MORE CHALLENGING PROBLEMS

Now that you can multiply fractions that have the number 1 as the numerator, you are ready to tackle these more complicated problems.

Problem: How much is three-fifths of three-quarters?

Solution: $\frac{3}{5} \times \frac{3}{4} = \frac{9}{20}$

Problem: How much is two-thirds of one-quarter?

Solution: $\frac{2}{3} \times \frac{1}{4} = \frac{2}{12} = \frac{1}{6}$

Problem: How much is a quarter of two-thirds?

Solution: $\frac{1}{4} \times \frac{2}{3} = \frac{2}{12} = \frac{1}{6}$

Did you notice what you just did in the last two problems? You just did the same problem and came up with the same answer. So two-thirds of one-quarter comes out the same as one-quarter of two-thirds. When you multiply proper fractions, you get the same answer regardless of the order in which you place the numbers. This is true of any type of multiplication problem.

PROBLEM SET

These problems are a bit more complicated than the first problem set in this lesson.

   7. How much is four-fifths of one-half?

   8. How much is nine-tenths of one-eighth?

   9. How much is four-sevenths of two-thirds?

   10. How much is eight-ninths of three-quarters?

   11. How much is five-sixths of four-fifths?

   12. How much is three-eighths of four-ninths?

Solutions

   7. $\frac{4}{5} \times \frac{1}{2} = \frac{4}{10} = \frac{2}{5}$

   8. $\frac{9}{10} \times \frac{1}{8} = \frac{9}{80}$

   9. $\frac{4}{7} \times \frac{2}{3} = \frac{8}{21}$

   10. $\frac{8}{9} \times \frac{3}{4} = \frac{24}{36} = \frac{2}{3}$

   11. $\frac{5}{6} \times \frac{4}{5} = \frac{20}{30} = \frac{2}{3}$

   12. $\frac{3}{8} \times \frac{4}{9} = \frac{12}{72} = \frac{1}{6}$

SHORTCUT: CANCELING OUT

When you multiply fractions, you can often save time and mental energy by *canceling out*. Here's how it works.

   Problem: How much is $\frac{5}{6} \times \frac{3}{4}$?

   Solution: $\frac{5}{6} \times \frac{3}{4} = \frac{5}{2\cancel{6}} \times \frac{\cancel{3}^{1}}{4} = \frac{5}{8}$

In this problem we performed a process called *canceling out*. We divided the 6 in $\frac{5}{6}$ by 3 and we divided the 3 in $\frac{3}{4}$ by 3. In other words, the 3 in the 6 and the 3 in the 3 canceled each other out. Try to cancel out the following problem.

Problem: How much is $\frac{2}{3} \times \frac{1}{2}$?

Solution: $\frac{\overset{1}{2}}{3} \times \frac{1}{\underset{1}{2}} = \frac{1}{3}$

Canceling out helps you reduce fractions to their lowest possible terms. While there's no law of arithmetic that says you have to do this, it makes multiplication easier, because it's much easier to work with smaller numbers. For example, suppose you had this problem:

Problem: $\frac{17}{20} \times \frac{5}{34} =$

Solution: There are two ways to solve this:

1. $\frac{17}{20} \times \frac{5}{34} = \frac{85}{680} = \frac{17}{136} = \frac{1}{8}$

2. $\frac{\overset{1}{17}}{\underset{4}{20}} \times \frac{\overset{1}{5}}{\underset{2}{34}} = \frac{1}{8}$

Obviously, the second version is much easier.

PROBLEM SET

In this problem set, see if you can cancel out before you multiply. If you're not comfortable doing this, then carry out the multiplication without canceling out. As long as you're getting the right answers, it doesn't matter whether or not you use this simplification tool.

13. $\frac{7}{16} \times \frac{8}{21} =$

14. $\frac{4}{9} \times \frac{3}{20} =$

15. $\frac{14}{32} \times \frac{16}{21} =$

16. $\frac{15}{24} \times \frac{9}{10} =$

17. $\frac{13}{20} \times \frac{5}{39} =$

18. $\frac{9}{42} \times \frac{7}{18} =$

Solutions

13. $\frac{\cancel{7}^{\,1}}{\cancel{16}_{2}} \times \frac{\cancel{8}^{\,1}}{\cancel{21}_{3}} = \frac{1}{6}$

14. $\frac{\cancel{4}^{\,1}}{\cancel{9}_{3}} \times \frac{\cancel{3}^{\,1}}{\cancel{20}_{5}} = \frac{1}{15}$

15. $\frac{\cancel{14}^{\,2}}{\cancel{32}_{2}} \times \frac{\cancel{16}^{\,1}}{\cancel{21}_{3}} = \frac{2}{6} = \frac{1}{3}$

16. $\frac{\cancel{15}^{\,3}}{\cancel{24}_{8}} \times \frac{\cancel{9}^{\,3}}{\cancel{10}_{2}} = \frac{9}{16}$

17. $\frac{\cancel{13}^{\,1}}{\cancel{20}_{4}} \times \frac{\cancel{5}^{\,1}}{\cancel{39}_{3}} = \frac{1}{12}$

18. $\frac{\cancel{9}^{\,1}}{\cancel{42}_{6}} \times \frac{\cancel{7}^{\,1}}{\cancel{18}_{2}} = \frac{1}{12}$

Next Step

So I didn't lie. Multiplying fractions is pretty easy. Dividing fractions, which we take up next, is virtually the same as multiplying fractions, except for one added step.

LESSON | 5

This lesson explains what a reciprocal is and shows you how to use it to solve fraction division problems. You'll also learn how to do division problems in the proper order so you get the right answer the first time.

DIVIDING FRACTIONS

The division of fractions is just like multiplication, but with a twist. You'll find the trick is to turn a division problem into a multiplication problem.

Let's get right into it. How much is one-third divided by one-half? Don't panic! The trick to doing this is to convert it into a multiplication problem. Just multiply one-third by the reciprocal of one-half. What did I say? The *reciprocal* of a fraction is found by turning the fraction upside down. So $\frac{1}{2}$ becomes $\frac{2}{1}$. With all this information, see if you can figure out the following problems.

Problem: How much is one-third divided by one-half?

Solution: $\frac{1}{3} \div \frac{1}{2} = \frac{1}{3} \times \frac{2}{1} = \frac{2}{3}$

Problem: How much is $\frac{1}{4}$ divided by $\frac{1}{6}$?

Solution: $\frac{1}{4} \div \frac{1}{6} = \frac{1}{2^4} \times \frac{\overset{3}{6}}{1} = \frac{3}{2} = 1\frac{1}{2}$

Let's just stop here for a minute. In the last problem we converted an improper fraction, $\frac{3}{2}$, into a mixed number, $1\frac{1}{2}$. I mentioned earlier how mathematicians just hate fractions that are not reduced to their lowest terms—$\frac{4}{6}$ must be reduced to $\frac{2}{3}$, and $\frac{4}{8}$ must be reduced to $\frac{1}{2}$. Another thing that really bothers them is leaving an improper fraction as an answer instead of converting it into a mixed number.

See if you can work out this problem:

Problem: $\frac{1}{3} \div \frac{1}{4} =$

Solution: $\frac{1}{3} \div \frac{1}{4} = \frac{1}{3} \times \frac{4}{1} = \frac{4}{3} = 1\frac{1}{3}$

Remember that, whenever you need to convert an improper fraction into a mixed number, you just divide the denominator (bottom number) into the numerator (top number). If you need to review this procedure, turn back to Lesson 1, near the beginning of this section.

PROBLEM SET
Divide each of these fractions by the one that follows.

1. $\frac{1}{2} \div \frac{1}{5} =$

2. $\frac{1}{8} \div \frac{1}{3} =$

3. $\frac{1}{9} \div \frac{1}{6} =$

4. $\frac{1}{4} \div \frac{1}{7} =$

5. $\frac{1}{6} \div \frac{1}{2} =$

6. $\frac{1}{3} \div \frac{1}{8} =$

Solutions

1. $\frac{1}{2} \div \frac{1}{5} = \frac{1}{2} \times \frac{5}{1} = \frac{5}{2} = 2\frac{1}{2}$

2. $\frac{1}{8} \div \frac{1}{3} = \frac{1}{8} \times \frac{3}{1} = \frac{3}{8}$

3. $\frac{1}{9} \div \frac{1}{6} = \frac{1}{\cancel{9}^3} \times \frac{\cancel{6}^2}{1} = \frac{2}{3}$

4. $\frac{1}{4} \div \frac{1}{7} = \frac{1}{4} \times \frac{7}{1} = \frac{7}{4} = 1\frac{3}{4}$

5. $\frac{1}{6} \div \frac{1}{2} = \frac{1}{\cancel{6}^3} \times \frac{\cancel{2}^1}{1} = \frac{1}{3}$

6. $\frac{1}{3} \div \frac{1}{8} = \frac{1}{3} \times \frac{8}{1} = \frac{8}{3} = 2\frac{2}{3}$

THE ORDER OF THE NUMBERS

When you multiply two numbers, you get the same answer regardless of their order. For example, $\frac{1}{2} \times \frac{1}{3}$ gives you the same answer as $\frac{1}{3} \times \frac{1}{2}$.

$$\frac{1}{2} \times \frac{1}{3} = \frac{1}{6}$$

$$\frac{1}{3} \times \frac{1}{2} = \frac{1}{6}$$

When you divide one number by another, does it matter in which order you write the numbers? Let's find out.

Problem: How much is $\frac{1}{3}$ divided by $\frac{1}{4}$?

Solution: $\frac{1}{3} \div \frac{1}{4} = \frac{1}{3} \times \frac{4}{1} = \frac{4}{3} = 1\frac{1}{3}$

Problem: Now how much is $\frac{1}{4}$ divided by $\frac{1}{3}$?

Solution: $\frac{1}{4} \div \frac{1}{3} = \frac{1}{4} \times \frac{3}{1} = \frac{3}{4}$

There is no way that $\frac{3}{4}$ can equal $1\frac{1}{3}$. So when you do division of fractions, you must be very careful about the order of the numbers. The number that is being divided always comes **before** the division sign, and the number doing the dividing always comes **after** the division sign. Here are some examples.

Problem: How much is $\frac{1}{5}$ divided by $\frac{1}{8}$?

Solution: $\frac{1}{5} \div \frac{1}{8} = \frac{1}{5} \times \frac{8}{1} = \frac{8}{5} = 1\frac{3}{5}$

Problem: How much is $\frac{1}{4}$ divided into $\frac{1}{7}$?

Solution: $\frac{1}{7} \div \frac{1}{4} = \frac{1}{7} \times \frac{4}{1} = \frac{4}{7}$

Problem: Moving right along, how much is $\frac{5}{6}$ divided by $\frac{2}{3}$?

Solution: $\frac{5}{6} \div \frac{2}{3} = \frac{5}{{}_2 6} \times \frac{3^1}{2} = \frac{5}{4} = 1\frac{1}{4}$

Problem: Divide $\frac{3}{5}$ by $\frac{3}{8}$.

Solution: $\frac{3}{5} \div \frac{3}{8} = \frac{3^1}{5} \times \frac{8}{{}_1 3} = \frac{8}{5} = 1\frac{3}{5}$

Problem: Now do this problem: $\frac{7}{10} \div \frac{3}{5} =$

Solution: $\frac{7}{10} \div \frac{3}{5} = \frac{7}{{}_2 10} \times \frac{5^1}{3} = \frac{7}{6} = 1\frac{1}{6}$

PROBLEM SET

Here's a problem set for you to complete.

6. Divide $\frac{3}{4}$ by $\frac{2}{3}$.

7. Divide $\frac{3}{5}$ by $\frac{3}{8}$.

8. How much is $\frac{5}{6}$ divided by $\frac{3}{4}$?

9. How much is $\frac{5}{8}$ divided by $\frac{5}{8}$?

10. How much is $\frac{2}{7}$ divided by $\frac{4}{5}$?

11. How much is $\frac{1}{4}$ divided by $\frac{5}{8}$?

Solutions

6. $\frac{3}{4} \div \frac{2}{3} = \frac{3}{4} \times \frac{3}{2} = \frac{9}{8} = 1\frac{1}{8}$

7. $\frac{3}{5} \div \frac{3}{8} = \frac{\cancel{3}^1}{5} \times \frac{8}{\cancel{3}_1} = \frac{8}{5} = 1\frac{3}{5}$

8. $\frac{5}{6} \div \frac{3}{4} = \frac{5}{\cancel{6}_3} \times \frac{\cancel{4}^2}{3} = \frac{10}{9} = 1\frac{1}{9}$

9. $\frac{5}{8} \div \frac{5}{8} = \frac{\cancel{5}^1}{\cancel{8}_1} \times \frac{\cancel{8}^1}{\cancel{5}_1} = \frac{1}{1} = 1$

10. $\frac{2}{7} \div \frac{4}{5} = \frac{\cancel{2}^1}{7} \times \frac{5}{\cancel{4}_2} = \frac{5}{14}$

11. $\frac{1}{4} \div \frac{5}{8} = \frac{1}{\cancel{4}_1} \times \frac{\cancel{8}^2}{5} = \frac{2}{5}$

MATHEMATICAL OBSERVATIONS

Take a look at the problem set you just did and make a few observations. In problem 9 you divided $\frac{5}{8}$ by $\frac{5}{8}$ and got an answer of 1. Any number divided by itself equals 1.

Problem: Try dividing $\frac{3}{4}$ by $\frac{3}{4}$.

Solution: $\frac{3}{4} \div \frac{3}{4} = \frac{\cancel{3}^1}{\cancel{4}_1} \times \frac{\cancel{4}^1}{\cancel{3}_1} = \frac{1}{1} = 1$

Next observation: In problem 6 you divided $\frac{3}{4}$ by a smaller number, $\frac{2}{3}$. Your answer was $1\frac{1}{8}$. In problem 7 we divided $\frac{3}{5}$ by a smaller number, $\frac{3}{8}$. Our answer was $1\frac{3}{5}$. You can generalize: When you divide a number by a smaller number, the answer (or quotient) will be greater than 1.

Final observation: In problem 10 you divided $\frac{2}{7}$ by a larger number, $\frac{4}{5}$. Your answer was $\frac{5}{14}$. In problem 11 you divided $\frac{1}{4}$ by a larger number, $\frac{5}{8}$. Your answer was $\frac{2}{5}$. Here's the generalization: When you divide a number by a larger number, the number (or quotient) will be less than 1.

NEXT STEP

At this point you should be able to add, subtract, multiply, and divide fractions. In the next lesson, we're going to throw it all at you at the same time.

LESSON | 6

In this lesson, you'll use everything you've learned so far in this entire section. So before going any further, make sure you know what you need to know about improper fractions and mixed numbers (Lesson 1), and about adding, subtracting, multiplying, and dividing proper fractions (Lessons 2, 3, 4, and 5, respectively).

WORKING WITH IMPROPER FRACTIONS

Do you really need to know how to add, subtract, multiply, and divide improper fractions? Yes! In the very next lesson, you'll need to convert mixed numbers into improper fractions before you can do addition, subtraction, multiplication, and division. I know that you can hardly wait.

ADDING WITH COMMON DENOMINATORS

Let's start by adding two improper fractions with common denominators.

Problem: How much is $\frac{12}{5} + \frac{7}{5}$?

Solution: $\frac{12}{5} + \frac{7}{5} = \frac{19}{5} = 3\frac{4}{5}$

You'll notice I converted the improper fraction, $\frac{19}{5}$, into a mixed number, $3\frac{4}{5}$. By convention, you should make this conversion with your answers.

Now try another problem.

Problem: $\frac{5}{3} + \frac{8}{3} + \frac{11}{3} =$

Solution: $\frac{5}{3} + \frac{8}{3} + \frac{11}{3} = \frac{24}{3} = 8$

PROBLEM SET
Add each set of fractions.

1. $\frac{4}{2} + \frac{7}{2} =$

2. $\frac{9}{3} + \frac{6}{3} =$

3. $\frac{23}{20} + \frac{39}{20} =$

4. $\frac{9}{4} + \frac{7}{4} + \frac{10}{4} =$

5. $\frac{11}{9} + \frac{10}{9} + \frac{15}{9} =$

6. $\frac{15}{7} + \frac{23}{7} + \frac{11}{7} =$

Solutions

1. $\frac{4}{2} + \frac{7}{2} = \frac{11}{2} = 5\frac{1}{2}$

2. $\frac{9}{3} + \frac{6}{3} = \frac{15}{3} = 5$

3. $\frac{23}{20} + \frac{39}{20} = \frac{62}{20} = 3\frac{2}{20} = 3\frac{1}{10}$

4. $\frac{9}{4} + \frac{7}{4} + \frac{10}{4} = \frac{26}{4} = 6\frac{2}{4} = 6\frac{1}{2}$

5. $\frac{11}{9} + \frac{10}{9} + \frac{15}{9} = \frac{36}{9} = 4$

6. $\frac{15}{7} + \frac{23}{7} + \frac{11}{7} = \frac{49}{7} = 7$

ADDING WITH UNLIKE DENOMINATORS

So far we've added improper fractions with common denominators. Here are a couple of fractions with different denominators to add.

Problem: $\frac{7}{2} + \frac{8}{3} =$

Solution: $\frac{7}{2} + \frac{8}{3} = \frac{7 \times 3}{2 \times 3} + \frac{8 \times 2}{3 \times 2} = \frac{21}{6} + \frac{16}{6} = \frac{37}{6} = 6\frac{1}{6}$

Problem: $\frac{14}{8} + \frac{10}{6} + \frac{7}{3} =$

Solution: $\frac{14}{8} + \frac{10}{6} + \frac{7}{3} = \frac{14 \times 3}{8 \times 3} + \frac{10 \times 4}{6 \times 4} + \frac{7 \times 8}{3 \times 8} = \frac{42}{24} + \frac{40}{24} + \frac{56}{24}$

$$= \frac{138}{24} = \frac{69}{12} = \frac{23}{4} = 5\frac{3}{4}$$

PROBLEM SET

Add each of these sets of fractions.

7. $\frac{12}{5} + \frac{13}{10} =$

8. $\frac{17}{9} + \frac{10}{3} =$

9. $\frac{20}{7} + \frac{9}{5} =$

10. $\frac{15}{4} + \frac{8}{5} + \frac{13}{6} =$

11. $\frac{16}{3} + \frac{21}{5} + \frac{17}{10} =$

12. $\frac{13}{8} + \frac{9}{4} + \frac{7}{2} =$

Solutions

7. $\frac{12}{5} + \frac{13}{10} = \frac{12 \times 2}{5 \times 2} + \frac{13}{10} = \frac{24}{10} + \frac{13}{10} = \frac{37}{10} = 3\frac{7}{10}$

8. $\frac{17}{9} + \frac{10}{3} = \frac{17}{9} + \frac{10 \times 3}{3 \times 3} = \frac{17}{9} + \frac{30}{9} = \frac{47}{9} = 5\frac{2}{9}$

9. $\frac{20}{7} + \frac{9}{5} = \frac{20 \times 5}{7 \times 5} + \frac{9 \times 7}{5 \times 7} = \frac{100}{35} + \frac{63}{35} = \frac{163}{35} = 4\frac{23}{35}$

10. $\frac{15}{4} + \frac{8}{5} + \frac{13}{6} = \frac{15 \times 15}{4 \times 15} + \frac{8 \times 12}{5 \times 12} + \frac{13 \times 10}{6 \times 10} = \frac{225}{60} + \frac{96}{60} + \frac{130}{60} = \frac{451}{60} = 7\frac{31}{60}$

11. $\frac{16}{3} + \frac{21}{5} + \frac{17}{10} = \frac{16 \times 10}{3 \times 10} + \frac{21 \times 6}{5 \times 6} + \frac{17 \times 3}{10 \times 3} = \frac{160}{30} + \frac{126}{30} + \frac{51}{30} = \frac{337}{30} = 11\frac{7}{30}$

12. $\frac{13}{8} + \frac{9}{4} + \frac{7}{2} = \frac{13}{8} + \frac{9 \times 2}{4 \times 2} + \frac{7 \times 4}{2 \times 4} = \frac{13}{8} + \frac{18}{8} + \frac{28}{8} = \frac{59}{8} = 7\frac{3}{8}$

Let's skip the second step of the solution to problem 12:

$$\frac{13}{8} + \frac{9}{4} + \frac{7}{2} = \frac{13}{8} + \frac{18}{8} + \frac{28}{8} = \frac{59}{8} = 7\frac{3}{8}$$

SUBTRACTING IMPROPER FRACTIONS

Are you ready for some subtraction?

Problem: How much is $\frac{15}{4} - \frac{11}{3}$?

Solution: $\frac{15}{4} - \frac{11}{3} = \frac{15 \times 3}{4 \times 3} - \frac{11 \times 4}{3 \times 4} = \frac{45}{12} - \frac{44}{12} = \frac{1}{12}$

Problem: How much is $\frac{24}{7} - \frac{19}{8}$?

Solution: $\frac{24}{7} - \frac{19}{8} = \frac{24 \times 8}{7 \times 8} - \frac{19 \times 7}{8 \times 7} = \frac{192}{56} - \frac{133}{56} = \frac{59}{56} = 1\frac{3}{56}$

PROBLEM SET
Try your hand at these subtraction problems.

13. $\frac{15}{2} - \frac{16}{3} =$

14. $\frac{22}{9} - \frac{4}{3} =$

15. $\frac{17}{5} - \frac{12}{10} =$

16. $\frac{15}{7} - \frac{5}{4} =$

17. $\frac{36}{8} - \frac{26}{10} =$

18. $\frac{59}{6} - \frac{49}{14} =$

Solutions

13. $\frac{15}{2} - \frac{16}{3} = \frac{15 \times 3}{2 \times 3} - \frac{16 \times 2}{3 \times 2} = \frac{45}{6} - \frac{32}{6} = \frac{13}{6} = 2\frac{1}{6}$

14. $\frac{22}{9} - \frac{4}{3} = \frac{22}{9} - \frac{4 \times 3}{3 \times 3} = \frac{22}{9} - \frac{12}{9} = \frac{10}{9} = 1\frac{1}{9}$

15. $\frac{17}{5} - \frac{12}{10} = \frac{17 \times 2}{5 \times 2} - \frac{12}{10} = \frac{34}{10} - \frac{12}{10} = \frac{22}{10} = 2\frac{2}{10} = 2\frac{1}{5}$

16. $\frac{15}{7} - \frac{5}{4} = \frac{15 \times 4}{7 \times 4} - \frac{5 \times 7}{4 \times 7} = \frac{60}{28} - \frac{35}{28} = \frac{25}{28}$

17. $\frac{36}{8} - \frac{26}{10} = \frac{36 \times 5}{8 \times 5} - \frac{26 \times 4}{10 \times 4} = \frac{180}{40} - \frac{104}{40} = \frac{76}{40} = \frac{19}{10} = 1\frac{9}{10}$

18. $\frac{59}{6} - \frac{49}{14} = \frac{59 \times 7}{6 \times 7} - \frac{49 \times 3}{14 \times 3} = \frac{413}{42} - \frac{147}{42} = \frac{266}{42} = \frac{133}{21} = 6\frac{7}{21} = 6\frac{1}{3}$

MULTIPLYING IMPROPER FRACTIONS

Multiplication of improper fractions is actually quite straightforward.

Problem: How much is $\frac{10}{3} \times \frac{12}{9}$?

Solution: $\frac{10}{{}_1 \cancel{3}} \times \frac{\cancel{12}^{\,4}}{9} = \frac{40}{9} = 4\frac{4}{9}$

PROBLEM SET

19. $\frac{17}{5} \times \frac{8}{3} =$

20. $\frac{23}{4} \times \frac{20}{9} =$

21. $\frac{21}{6} \times \frac{12}{7} =$

22. $\frac{15}{8} \times \frac{16}{5} =$

23. $\frac{46}{9} \times \frac{13}{4} =$

24. $\frac{17}{10} \times \frac{25}{8} =$

Solutions

19. $\frac{17}{5} \times \frac{8}{3} = \frac{136}{15} = 9\frac{1}{15}$

20. $\frac{23}{1\,4} \times \frac{\cancel{20}^{5}}{9} = \frac{115}{9} = 12\frac{7}{9}$

21. $\frac{\cancel{21}^{3}}{1\,6} \times \frac{\cancel{12}^{2}}{1\,7} = \frac{6}{1} = 6$

22. $\frac{\cancel{15}^{3}}{1\,8} \times \frac{\cancel{16}^{2}}{1\,5} = \frac{6}{1} = 6$

23. $\frac{\cancel{46}^{23}}{9} \times \frac{13}{2\,4} = \frac{299}{18} = 16\frac{11}{18}$

24. $\frac{17}{2\,\cancel{10}} \times \frac{\cancel{25}^{5}}{8} = \frac{85}{16} = 5\frac{5}{16}$

DIVIDING IMPROPER FRACTIONS

Like multiplication of improper fractions, division is also quite straight-forward.

Problem: How much is $\frac{14}{9}$ divided by $\frac{10}{3}$?

Solution: $\frac{14}{9} \div \frac{10}{3} = \frac{\cancel{14}^{7}}{3\,9} \times \frac{\cancel{3}^{1}}{5\,\cancel{10}} = \frac{7}{15}$

PROBLEM SET

Complete these division problems.

25. $\frac{32}{5} \div \frac{18}{15} =$

26. $\frac{17}{4} \div \frac{9}{2} =$

27. $\frac{25}{7} \div \frac{15}{14} =$

28. $\frac{31}{6} \div \frac{3}{2} =$

29 $\frac{72}{11} \div \frac{6}{5} =$

30. $\frac{46}{13} \div \frac{15}{13} =$

Solutions

25. $\frac{32}{5} \div \frac{18}{15} = {}_{1}\frac{\cancel{32}}{\cancel{5}}^{16} \times {}_{9}\frac{\cancel{15}}{\cancel{18}}^{3} = \frac{48}{9} = 5\frac{3}{9} = 5\frac{1}{3}$

26. $\frac{17}{4} \div \frac{9}{2} = {}_{2}\frac{17}{\cancel{4}} \times \frac{\cancel{2}}{9}^{1} = \frac{17}{18}$

27. $\frac{25}{7} \div \frac{15}{14} = {}_{1}\frac{\cancel{25}}{\cancel{7}}^{5} \times {}_{3}\frac{\cancel{14}}{\cancel{15}}^{2} = \frac{10}{3} = 3\frac{1}{3}$

28. $\frac{31}{6} \div \frac{3}{2} = {}_{3}\frac{31}{\cancel{6}} \times \frac{\cancel{2}}{3}^{1} = \frac{31}{9} = 3\frac{4}{9}$

29. $\frac{72}{11} \div \frac{6}{5} = \frac{\cancel{72}}{11}^{12} \times {}_{1}\frac{5}{\cancel{6}} = \frac{60}{11} = 5\frac{5}{11}$

30. $\frac{46}{13} \div \frac{15}{13} = {}_{1}\frac{46}{\cancel{13}} \times \frac{\cancel{13}}{15}^{1} = \frac{46}{15} = 3\frac{1}{15}$

NEXT STEP

Now that you know how to add, subtract, multiply, and divide improper fractions, you'll be using that skill to perform the same tricks with mixed numbers. The only additional trick you'll need to do is to convert mixed numbers into improper fractions and improper fractions into mixed numbers. If you don't remember how to do this, you'll need to go back and look at Lesson 1 again.

LESSON | 7

WORKING WITH MIXED NUMBERS

Remember mixed numbers? (Right, a mixed number is a whole number plus a fraction.) This lesson will show you how to add, subtract, multiply, and divide mixed numbers. You'll have to convert mixed numbers into improper fractions first, so make sure you're up on your "Fraction Conversions" (Lesson 1).

In this lesson you'll put together everything you've learned so far about fractions. In order to perform operations on mixed numbers, you'll be following a three-step process:

1. Convert mixed numbers into improper fractions.
2. Add, subtract, multiply, or divide.
3. Convert improper fractions into mixed numbers.

You'll add one step here to what you did in the previous lesson. Before you can add, subtract, multiply, or divide mixed numbers, you need to convert them into improper fractions. Once you've done that, you can do exactly what you did in Lesson 6.

ADDING MIXED NUMBERS

Here's a problem to get you started.

Problem: Add $2\frac{5}{8}$ and $1\frac{3}{4}$.

Solution: $2\frac{5}{8} + 1\frac{3}{4} = \frac{21}{8} + \frac{7}{4} = \frac{21}{8} + \frac{7\times2}{4\times2} = \frac{21}{8} + \frac{14}{8} = \frac{35}{8} = 4\frac{3}{8}$

PROBLEM SET

Add each of these sets of mixed numbers.

1. $3\frac{4}{5} + 2\frac{7}{8} =$

2. $5\frac{2}{7} + 3\frac{5}{8} =$

3. $2\frac{1}{8} + 5\frac{2}{3} =$

4. $7\frac{5}{9} + 4\frac{1}{3} =$

5. $1\frac{3}{4} + 2\frac{7}{8} + 3\frac{1}{2} =$

6. $4\frac{3}{7} + 2\frac{1}{6} + 3\frac{4}{5} =$

Solutions

1. $3\frac{4}{5} + 2\frac{7}{8} = \frac{19}{5} + \frac{23}{8} = \frac{19\times8}{5\times8} + \frac{23\times5}{8\times5} = \frac{152}{40} + \frac{115}{40} = \frac{267}{40} = 6\frac{27}{40}$

2. $5\frac{2}{7} + 3\frac{5}{8} = \frac{37}{7} + \frac{29}{8} = \frac{37\times8}{7\times8} + \frac{29\times7}{8\times7} = \frac{296}{56} + \frac{203}{56} = \frac{499}{56} = 8\frac{51}{56}$

3. $2\frac{1}{8} + 5\frac{2}{3} = \frac{17}{8} + \frac{17}{3} = \frac{17\times3}{8\times3} + \frac{17\times8}{3\times8} = \frac{51}{24} + \frac{136}{24} = \frac{187}{24} = 7\frac{19}{24}$

4. $7\frac{5}{9} + 4\frac{1}{3} = \frac{68}{9} + \frac{13}{3} = \frac{68}{9} + \frac{13\times3}{3\times3} = \frac{68}{9} + \frac{39}{9} = \frac{107}{9} = 11\frac{8}{9}$

5. $1\frac{3}{4} + 2\frac{7}{8} + 3\frac{1}{2} = \frac{7}{4} + \frac{23}{8} + \frac{7}{2} = \frac{7\times2}{4\times2} + \frac{23}{8} + \frac{7\times4}{2\times4} = \frac{14}{8} + \frac{23}{8} + \frac{28}{8}$

    $= \frac{65}{8} = 8\frac{1}{8}$

6. $4\frac{3}{7} + 2\frac{1}{6} + 3\frac{4}{5} = \frac{31}{7} + \frac{13}{6} + \frac{19}{5} = \frac{31 \times 30}{7 \times 30} + \frac{13 \times 35}{6 \times 35} + \frac{19 \times 42}{5 \times 42}$

$= \frac{930}{210} + \frac{455}{210} + \frac{798}{210} = \frac{2183}{210} = 10\frac{83}{210}$

It might have occurred to you that there's another way to do these problems. You could add whole numbers, add fractions, and then add them together, *carrying* where necessary. In other words, you could do problem 1 like this:

Problem: $3\frac{4}{5} + 2\frac{7}{8} =$

Solution: $3 + 2 + \frac{4}{5} + \frac{7}{8} = 5 + \frac{32}{40} + \frac{35}{40} = 5\frac{67}{40} = 6\frac{27}{40}$

The problem with this method is that you might forget to carry. So stick with my method of converting to improper fractions.

SUBTRACTING MIXED NUMBERS

Now it's time to subtract mixed numbers.

Problem: Find the answer to $4\frac{5}{7} - 2\frac{3}{4}$.

Solution: $4\frac{5}{7} - 2\frac{3}{4} = \frac{33}{7} - \frac{11}{4} = \frac{33 \times 4}{7 \times 4} - \frac{11 \times 7}{4 \times 7} = \frac{132}{28} - \frac{77}{28} = \frac{55}{28} = 1\frac{27}{28}$

PROBLEM SET

Here's another problem set for you to complete.

7. $3\frac{1}{3} - 1\frac{5}{8} =$

8. $5\frac{4}{9} - 2\frac{2}{3} =$

9. $4\frac{3}{8} - 2\frac{3}{5} =$

10. $7\frac{5}{12} - 4\frac{5}{6} =$

11. $6\frac{3}{4} - 3\frac{2}{3} =$

12. $9\frac{1}{6} - 2\frac{3}{10} =$

Solutions

7. $3\frac{1}{3} - 1\frac{5}{8} = \frac{10}{3} - \frac{13}{8} = \frac{10 \times 8}{3 \times 8} - \frac{13 \times 3}{8 \times 3} = \frac{80}{24} - \frac{39}{24} = \frac{41}{24} = 1\frac{17}{24}$

8. $5\frac{4}{9} - 2\frac{2}{3} = \frac{49}{9} - \frac{8}{3} = \frac{49}{9} - \frac{8 \times 3}{3 \times 3} = \frac{49}{9} - \frac{24}{9} = \frac{25}{9} = 2\frac{7}{9}$

9. $4\frac{3}{8} - 2\frac{3}{5} = \frac{35}{8} - \frac{13}{5} = \frac{35 \times 5}{8 \times 5} - \frac{13 \times 8}{5 \times 8} = \frac{175}{40} - \frac{104}{40} = \frac{71}{40} = 1\frac{31}{40}$

10. $7\frac{5}{12} - 4\frac{5}{6} = \frac{89}{12} - \frac{29}{6} = \frac{89}{12} - \frac{29 \times 2}{6 \times 2} = \frac{89}{12} - \frac{58}{12} = \frac{31}{12} = 2\frac{7}{12}$

11. $6\frac{3}{4} - 3\frac{2}{3} = \frac{27}{4} - \frac{11}{3} = \frac{27 \times 3}{4 \times 3} - \frac{11 \times 4}{3 \times 4} = \frac{81}{12} - \frac{44}{12} = \frac{37}{12} = 3\frac{1}{12}$

12. $9\frac{1}{6} - 2\frac{3}{10} = \frac{55}{6} - \frac{23}{10} = \frac{55 \times 5}{6 \times 5} - \frac{23 \times 3}{10 \times 3} = \frac{275}{30} - \frac{69}{30} = \frac{206}{30} = \frac{103}{15} = 6\frac{13}{15}$

Sometimes we don't need to know the exact answer. All we really need is a fast estimate. In problem 10, we can quickly estimate our answer as between 2 and 3. In problem 11, our answer will be just a bit over 3. And in problem 12, the answer is going to be a little less than 7.

MULTIPLYING MIXED NUMBERS

Now you're ready to multiply mixed numbers.

Problem: How much is $3\frac{2}{3} \times 2\frac{1}{4}$?

Solution: $3\frac{2}{3} \times 2\frac{1}{4} = \frac{11}{3} \times \frac{9}{4} = \frac{99}{12} = \frac{33}{4} = 8\frac{1}{4}$

You may also cancel out to get the answer: $\frac{11}{{}_13} \times \frac{9{}^3}{4} = \frac{33}{4} = 8\frac{1}{4}$

PROBLEM SET

Do the following multiplication problems.

13. $1\frac{2}{3} \times 2\frac{1}{2} =$

14. $4\frac{1}{8} \times 2\frac{2}{5} =$

15. $5\frac{7}{8} \times 3\frac{1}{3} =$

16. $2\frac{4}{7} \times 4\frac{1}{9} =$

17. $3\frac{1}{2} \times 6\frac{2}{5} =$

18. $1\frac{8}{9} \times 4\frac{2}{3} =$

Solutions

13. $1\frac{2}{3} \times 2\frac{1}{2} = \frac{5}{3} \times \frac{5}{2} = \frac{25}{6} = 4\frac{1}{6}$

14. $4\frac{1}{8} \times 2\frac{2}{5} = \frac{33}{8} \times \frac{12}{5} = \frac{396}{40} = 9\frac{36}{40} = 9\frac{9}{10}$

15. $5\frac{7}{8} \times 3\frac{1}{3} = \frac{47}{8} \times \frac{10}{3} = \frac{470}{24} = \frac{235}{12} = 19\frac{7}{12}$

16. $2\frac{4}{7} \times 4\frac{1}{9} = \frac{\overset{2}{\cancel{18}}}{7} \times \frac{37}{\underset{1}{\cancel{9}}} = \frac{74}{7} = 10\frac{4}{7}$

17. $3\frac{1}{2} \times 6\frac{2}{5} = \frac{7}{\underset{1}{\cancel{2}}} \times \frac{\overset{16}{\cancel{32}}}{5} = \frac{112}{5} = 22\frac{2}{5}$

18. $1\frac{8}{9} \times 4\frac{2}{3} = \frac{17}{9} \times \frac{14}{3} = \frac{238}{27} = 8\frac{22}{27}$

DIVIDING MIXED NUMBERS

Finally we come to division of mixed numbers.

Problem: Divide $2\frac{6}{7}$ by $1\frac{3}{5}$.

Solution: $2\frac{6}{7} \div 1\frac{3}{5} = \frac{20}{7} \div \frac{8}{5} = \frac{\overset{5}{\cancel{20}}}{7} \times \frac{5}{\underset{2}{\cancel{8}}} = \frac{25}{14} = 1\frac{11}{14}$

Problem Set

Do each of these division problems.

19. $1\frac{1}{2} \div 2\frac{1}{3} =$

20. $3\frac{3}{4} \div 5\frac{7}{8} =$

21. $1\frac{1}{9} \div 5\frac{4}{5} =$

22. $5\frac{2}{3} \div 4\frac{1}{8} =$

23. $6\frac{1}{4} \div 3\frac{7}{9} =$

24. $2\frac{1}{2} \div 5\frac{6}{7} =$

Solutions

19. $1\frac{1}{2} \div 2\frac{1}{3} = \frac{3}{2} \div \frac{7}{3} = \frac{3}{2} \times \frac{3}{7} = \frac{9}{14}$

20. $3\frac{3}{4} \div 5\frac{7}{8} = \frac{15}{4} \div \frac{47}{8} = \frac{15}{{}_1 4} \times \frac{\cancel{8}^{\,2}}{47} = \frac{30}{47}$

21. $1\frac{1}{9} \div 5\frac{4}{5} = \frac{10}{9} \div \frac{29}{5} = \frac{10}{9} \times \frac{5}{29} = \frac{50}{261}$

22. $5\frac{2}{3} \div 4\frac{1}{8} = \frac{17}{3} \div \frac{33}{8} = \frac{17}{3} \times \frac{8}{33} = \frac{136}{99} = 1\frac{37}{99}$

23. $6\frac{1}{4} \div 3\frac{7}{9} = \frac{25}{4} \div \frac{34}{9} = \frac{25}{4} \times \frac{9}{34} = \frac{225}{136} = 1\frac{89}{136}$

24. $2\frac{1}{2} \div 5\frac{6}{7} = \frac{5}{2} \div \frac{41}{7} = \frac{5}{2} \times \frac{7}{41} = \frac{35}{82}$

NEXT STEP

You may have been doing most or all of these problems more or less mechanically. In the next lesson, you're going to have to think before you add, subtract, multiply, or divide. In fact, you're going to have to think about *whether* you're going to add, subtract, multiply, or divide. You'll get to do this by applying everything that you have learned so far.

LESSON | 8

This lesson gives you the opportunity to put your fraction knowledge to work. It's now time to apply everything you've learned so far in this section to real-world problems. Consider this lesson to be a practical application of all the principles you've learned in the fractions section.

APPLICATIONS

While no new material will be covered in this lesson, the math problems will be stated in words, and you'll need to translate these words into addition, subtraction, multiplication, and division problems, which you'll then solve. In the mathematical world, this type of math problem is often called a *word problem*.

PROBLEM SET

Do all of the problems on the next pages, and then check your work with the solutions that follow.

1. One morning you walked $4\frac{7}{8}$ miles to town. On the way home, you stopped to rest after walking $1\frac{1}{3}$ miles. How far do you still need to walk to get home?

2. To do an experiment, Sam needed $\frac{1}{12}$ of a gram of cobalt. If Eileen gave him $\frac{1}{4}$ of that amount, how much cobalt did she give Sam?

3. In an election, the Conservative candidate got one-eighth of the votes, the Republican candidate got one-sixth of the votes, and the Democratic candidate got one-third of the votes. What fraction of the votes did the three candidates receive all together?

4. When old man Jones died, his will left two-thirds of his fortune to his four children, and instructed them to divide their inheritance equally. What share of his fortune did each of his children receive?

5. Kerry is 4 feet $4\frac{1}{4}$ inches tall, and Mark is 4 feet $2\frac{7}{8}$ inches tall. How much taller is Kerry than Mark?

6. If you want to fence in your square yard, how much fencing would you need if your yard is $21\frac{2}{3}$ feet long? Remember that a square has four equal sides.

7. Ben and seven other friends bought a quarter share of a restaurant chain. If they were equal partners, what fraction of the restaurant chain did Ben own?

8. If it rained $1\frac{1}{2}$ inches on Monday, $2\frac{1}{8}$ inches on Tuesday, $\frac{3}{4}$ of an inch on Wednesday, and $2\frac{5}{8}$ inches on Thursday, how much did it rain over the four-day period?

9. If four and a half slices of pizza were divided equally among six people, how much pizza does each person get?

10. If four and three-quarter pounds of sand can fit in a box, how many pounds of sand can fit in six and a half boxes?

11. Ben Wallach opened a quart of orange juice in the morning. If he drank $\frac{1}{5}$ of it with breakfast and $\frac{2}{7}$ of it with lunch, how much of it did he have left for the rest of the day?

12. If Kit Hawkins bought $\frac{1}{3}$ of $\frac{1}{8}$ share in a company, what fraction of the company did she own?

13. At Elizabeth Zimiles' birthday party, there were four cakes. Each guest ate $\frac{1}{8}$ of a cake. How much cake was left over if there were 20 guests?

14. Suppose it takes $2\frac{1}{4}$ yards of material to make one dress. How many dresses could be made from a 900-yard bolt of material?

15. Max Krauthammer went on a diet and lost $4\frac{1}{2}$ pounds the first week, $3\frac{1}{2}$ the second week, $3\frac{1}{4}$ the third week, and $2\frac{3}{4}$ the fourth week. How much weight did he lose during the four weeks he dieted?

16. Sam Retchnick is a civil servant. He earns a half day of vacation time for every two weeks of work. How much vacation time does he earn for working $6\frac{1}{2}$ weeks?

17. Goodman Klang has been steadily losing $1\frac{1}{2}$ pounds a week on his diet. How much weight would he lose in $10\frac{1}{2}$ weeks?

18. Karen, Jeff, and Sophie pulled an all-nighter before an exam. They ordered 2 large pizzas and finished them by daybreak. If Karen had $\frac{2}{3}$ of a pie and Jeff had $\frac{3}{4}$, how much did Sophie have?

19. If four dogs split $6\frac{1}{2}$ cans of dog food equally, how much would each dog eat?

20. If Jason worked $9\frac{1}{2}$ hours on Monday, $8\frac{1}{4}$ hours on Tuesday, $7\frac{3}{4}$ hours on Wednesday, 9 hours on Thursday, and took Friday off, how many hours did he work that week?

Solutions

1. $4\frac{7}{8} - 1\frac{1}{3} = \frac{39}{8} - \frac{4}{3} = \frac{39 \times 3}{8 \times 3} - \frac{4 \times 8}{3 \times 8} = \frac{117}{24} - \frac{32}{24} = \frac{85}{24} = 3\frac{13}{24}$ miles

2. $\frac{1}{12} \times \frac{1}{4} = \frac{1}{48}$ of a gram

3. $\frac{1}{8} + \frac{1}{6} + \frac{1}{3} = \frac{1 \times 3}{8 \times 3} + \frac{1 \times 4}{6 \times 4} + \frac{1 \times 8}{3 \times 8} = \frac{3}{24} + \frac{4}{24} + \frac{8}{24} = \frac{15}{24} = \frac{5}{8}$ of the votes

4. $\frac{2}{3} \div \frac{4}{1} = \frac{\cancel{2}^1}{3} \times \frac{1}{\cancel{4}_2} = \frac{1}{6}$ of the fortune

5. $4\frac{1}{4} - 2\frac{7}{8} = \frac{17}{4} - \frac{23}{8} = \frac{17 \times 2}{4 \times 2} - \frac{23}{8} = \frac{34}{8} - \frac{23}{8} = \frac{11}{8} = 1\frac{3}{8}$ inches

6. $4 \times 21\frac{2}{3} = \frac{4}{1} \times \frac{65}{3} = \frac{260}{3} = 86\frac{2}{3}$ feet

7. $\frac{1}{4} \times \frac{1}{8} = \frac{1}{32}$ of the restaurant chain

8. $1\frac{1}{2} + 2\frac{1}{8} + \frac{3}{4} + 2\frac{5}{8} = \frac{3}{2} + \frac{17}{8} + \frac{3}{4} + \frac{21}{8} = \frac{3 \times 4}{2 \times 4} + \frac{17}{8} + \frac{3 \times 2}{4 \times 2} + \frac{21}{8}$

$= \frac{12}{8} + \frac{17}{8} + \frac{6}{8} + \frac{21}{8} = \frac{56}{8} = 7$ inches

9. $4\frac{1}{2} \div 6 = \frac{9}{2} \div \frac{6}{1} = \frac{\cancel{9}^3}{2} \times \frac{1}{\cancel{6}_2} = \frac{3}{4}$ of a slice

10. $4\frac{3}{4} \times 6\frac{1}{2} = \frac{19}{4} \times \frac{13}{2} = \frac{247}{8} = 30\frac{7}{8}$ pounds of sand

11. $1 - (\frac{1}{5} + \frac{2}{7}) = 1 - (\frac{1 \times 7}{5 \times 7} + \frac{2 \times 5}{7 \times 5}) = 1 - (\frac{7}{35} + \frac{10}{35}) = 1 - \frac{17}{35}$

$= \frac{35}{35} - \frac{17}{35} = \frac{18}{35}$ quart

12. $\frac{1}{3} \times \frac{1}{8} = \frac{1}{24}$ of the company

13. $4 - (20 \times \frac{1}{8}) = 4 - (\frac{20}{1} \times \frac{1}{8}) = 4 - \frac{20}{8} = \frac{32}{8} - \frac{20}{8} = \frac{12}{8} = 1\frac{4}{8} = 1\frac{1}{2}$ cakes

14. $900 \div 2\frac{1}{4} = \frac{900}{1} \div \frac{9}{4} = \frac{\cancel{900}^{100}}{1} \times \frac{4}{\cancel{9}_1} = 400$ dresses

15. $4\frac{1}{2} + 3\frac{1}{2} + 3\frac{1}{4} + 2\frac{3}{4} = \frac{9}{2} + \frac{7}{2} + \frac{13}{4} + \frac{11}{4} = \frac{9 \times 2}{2 \times 2} + \frac{7 \times 2}{2 \times 2} + \frac{13}{4} + \frac{11}{4}$

$= \frac{18}{4} + \frac{14}{4} + \frac{13}{4} + \frac{11}{4} = \frac{56}{4} = 14$ pounds

16. If Sam earns $\frac{1}{2}$ day for 2 weeks, then he earns $\frac{1}{4}$ day for one week.

$$\frac{1}{4} \times 6\frac{1}{2} = \frac{1}{4} \times \frac{13}{2} = \frac{13}{8} = 1\frac{5}{8} \text{ of a day}$$

17. $1\frac{1}{2} \times 10\frac{1}{2} = \frac{3}{2} \times \frac{21}{2} = \frac{63}{4} = 15\frac{3}{4}$ pounds

18. $2 - \left(\frac{2}{3} + \frac{3}{4}\right) = 2 - \left(\frac{2 \times 4}{3 \times 4} + \frac{3 \times 3}{4 \times 3}\right) = 2 - \left(\frac{8}{12} + \frac{9}{12}\right) = 2 - \frac{17}{12} = \frac{2 \times 12}{1 \times 12} - \frac{17}{12}$

$$= \frac{24}{12} - \frac{17}{12} = \frac{7}{12} \text{ of a pie}$$

19. $6\frac{1}{2} \div 4 = \frac{13}{2} \div \frac{4}{1} = \frac{13}{2} \times \frac{1}{4} = \frac{13}{8} = 1\frac{5}{8}$ cans

20. $9\frac{1}{2} + 8\frac{1}{4} + 7\frac{3}{4} + 9 = \frac{19}{2} + \frac{33}{4} + \frac{31}{4} + 9 = \frac{19 \times 2}{2 \times 2} + \frac{33}{4} + \frac{31}{4} + \frac{36}{4}$

$$= \frac{38}{4} + \frac{33}{4} + \frac{31}{4} + \frac{36}{4} = \frac{138}{4} = 34\frac{2}{4} \text{ hours}$$

If you bought 100 shares of Microsoft at $109\frac{3}{4}$ and sold them at $116\frac{3}{8}$, how much profit would you have made? (Don't worry about paying stockbrokers' commissions.)

Solution

When a stock has a price of $109\frac{3}{4}$, it is selling at \$109.75, or 109 and $\frac{3}{4}$ dollars. A fast way of working out this problem is to first look at the difference between $109\frac{3}{4}$ and $116\frac{3}{8}$. Let's ask ourselves the question, how much is $16\frac{3}{8} - 9\frac{3}{4}$? (We'll add on the 100 later.)

$$16\frac{3}{8} - 9\frac{3}{4} = \frac{131}{8} - \frac{39}{4} = \frac{131}{8} - \frac{78}{8} = \frac{53}{8} = 6\frac{5}{8}, \text{ or } \$6.625.$$

That's the profit you made on one share. Since you bought and sold 100 shares, you made a profit of \$662.50. This problem could also be worked out with decimals, which we'll do at the end of Lesson 13.

NEXT STEP

How are you doing so far? If you're getting everything right, or maybe just making a mistake here and there, then you're definitely ready for the next section. Two of the things you'll be doing are converting fractions into decimals and decimals into fractions. So before you start the next section, you need to be sure that you really have your fractions down cold. If you'd be more comfortable reviewing some or all of the work in this section, please allow yourself the time to do so.

SECTION | III

DECIMALS

What's a decimal? Like a fraction, a decimal is a part of one. One-half, or $\frac{1}{2}$, can be written as the decimal 0.5. By convention, decimals of less than 1 are preceded by 0.

Now let's talk about the decimal 0.1, which can be expressed as one-tenth, or $\frac{1}{10}$. Every decimal has a fractional equivalent and vice versa. And as you'll discover in this section, fractions and decimals also have percent equivalents.

Later in the section, you'll be converting tenths, hundredths, and thousandths from fractions into decimals and from decimals into fractions. And believe it or not, you'll be able to do all of this without even using a calculator.

When you have completed this lesson, you will know how to add, subtract, multiply, and divide decimals and convert fractions into decimals and decimals into fractions. You'll also see that the dollar is based on fractions and decimals.

LESSON | 9

In this lesson, you'll learn how to add and subtract numbers that are decimals. You'll also discover the importance of lining up the decimal points correctly before you begin to work a decimal problem.

ADDING AND SUBTRACTING DECIMALS

I f you spent $4.35 for a sandwich and $0.75 for a soda, how much did you spend for lunch? That's a decimal addition problem. If you had $24.36 in your pocket before lunch, how much did you have left after lunch? That's a decimal subtraction problem. Adding and subtracting decimals is just everyday math.

When you're adding and subtracting decimals, mathematically speaking, you're carrying out the same operations as when you're adding and subtracting whole numbers. Just keep your columns straight and keep track of where you're placing the decimal in your answers.

ADDING DECIMALS

Remember to be careful about lining up decimal points when adding decimals. These first problems are quite straightforward.

Problem: 1.96
              + 4.75

Solution: $^{1\ 1}$1.96
              + 4.75
               6.71

Now let's do one that's a little longer:

Problem: 2.83
              7.06
              5.14
              + 3.92

Solution: $^{1\ 1}$2.83
              7.06
              5.14
              + 3.92
              18.95

Problem: Suppose you drove across the country in six days. How much was your total mileage if you went these distances: 462.3 miles, 507.1 miles, 482.0 miles, 466.5 miles, 510.8 miles, and 495.3 miles?

Solution: $^{3\ 2\ 2}$462.3
              507.1
              482.0
              466.5
              510.8
              + 495.3
              2,924.0

Problem: It rained every day for the last week. You need to find the total rainfall for the week. Here's the recorded rainfall: Sunday, 1.22 inches; Monday, 0.13 inches; Tuesday, 2.09 inches; Wednesday, 0.34 inches; Thursday, 0.26 inches; Friday, 1.88 inches; and Saturday, 2.74 inches.

Solution:

$$
\begin{array}{r}
\overset{2}{} \overset{3}{} \\
1.22 \\
0.13 \\
2.09 \\
0.34 \\
0.26 \\
1.88 \\
+\ 2.74 \\
\hline
8.66
\end{array}
$$

In this last problem, you probably noticed the recorded rainfall for Monday (0.13), Wednesday (0.34), and Thursday (0.26) began with a zero. Do you have to place a zero in front of a decimal point? No, but when you're adding these decimals with other decimals that have values of more than 1, placing a zero in front of the decimal point not only helps you keep your columns straight, but it also helps prevent mistakes. Here's a set of problems to work out.

PROBLEM SET

Add each of these sets of numbers. Two sets are printed across, so you can practice aligning the decimal points in the correct columns.

1.		2.	
	4.5		513.38
	17.33		469.01
	9.01		137.59
	2.0		12.0
	+ 7.9		+ 173.09

3. $160.81 + 238.5 + 79.43 + 63.0 + 15.72 =$

4. $3.02 + 7.4 + 19.56 + 43.75 =$

Solutions

1.

$$\overset{1}{4}.5$$
$$\overset{3}{17}.33$$
$$9.01$$
$$2.0$$
$$\underline{+\ 7.9}$$
$$40.74$$

3.

$$\overset{2\,2\,2}{160}.81$$
$$238.5$$
$$79.43$$
$$63.0$$
$$\underline{+\ 15.72}$$
$$557.46$$

2.

$$\overset{2\,2\,1\,2}{513}.38$$
$$469.01$$
$$137.59$$
$$12.0$$
$$\underline{+\ 173.09}$$
$$1,305.07$$

4.

$$\overset{1\ 1}{3}.02$$
$$7.4$$
$$\overset{2}{19}.56$$
$$\underline{+\ 43.75}$$
$$73.73$$

SUBTRACTING DECIMALS

Are you ready for some subtraction? Subtracting decimals can be almost as much fun as adding them. See what you can do with this one.

Problem:

$$4.33$$
$$\underline{-\ 2.56}$$

Solution:

$$\overset{3\ 12\ 1}{\cancel{4.3}3}$$
$$\underline{-\ 2.56}$$
$$1.77$$

Let's try another one.

Problem:

$$30.41$$
$$\underline{-\ 19.73}$$

Solution:

$$\overset{2\ 9\ 13\ 1}{\cancel{30.4}1}$$
$$\underline{-\ 19.73}$$
$$10.68$$

Here come a couple of word problems.

Problem: Roberto weighs 113.2 pounds, and Melissa weighs 88.4 pounds. How much more than Melissa does Roberto weigh?

Solution:

$$
\begin{array}{r}
\overset{10\ 12\ 1}{\cancel{113}.2} \\
-\ 88.4 \\
\hline
24.8
\end{array}
$$

Problem: The population of Mexico is 78.79 million, and the population of the United States is 270.4 million. How many more people live in the United States than in Mexico?

Solution:

$$
\begin{array}{r}
\overset{1\ 16\ 9\ 13\ 1}{\cancel{270.40}} \\
-\ 78.79 \\
\hline
191.61
\end{array}
$$

That was a bit of a trick question. I wanted you to add a zero to the 270.4 million population of the United States. Why? To make the subtraction easier and to help you get the right answer. Adding the zero makes it easier to line up the decimal points—and you *have* to line up the decimal points to get the right answer.

You are allowed to add zeros to the right of decimals. You could have made 270.4 into 270.40000 if you wished. The only reason you add zeros is to help you line up the decimal points when you do addition or subtraction. Let's try one more.

Problem: Kevin scored 9.042 in gymnastics competition, but 0.15 points were deducted from his score for wearing the wrong sneakers. How much was his corrected, or lowered, score?

Solution:

$$
\begin{array}{r}
\overset{8\ 9\ 1}{\cancel{9.0}42} \\
-\ 0.150 \\
\hline
8.892
\end{array}
$$

Again, you added a 0 after 0.150 so you could line it up with 9.042 easily.

PROBLEM SET

Carry out each of these subtraction problems. You'll have to line up the last problem yourself.

5.   121.06
          − 98.34

7.   812.71
          − 626.78

6.   709.44
          − 529.65

8. Subtract 39.48 from 54.35.

Solutions

5. $\overset{11\ 10\ \ 1}{1\cancel{21}.06}$
          − 9 8.34
          2 2.72

7. $\overset{7\ 10\ 11\ 16\ 1}{8\cancel{12}.\cancel{7}1}$
          − 6 2 6.78
          1 8 5.93

6. $\overset{6\ 9\ 18\ 13\ 1}{\cancel{709}.44}$
          − 5 2 9.65
          1 7 9.79

8. $\overset{4\ 13\ 12\ 1}{5\cancel{4}.\cancel{3}5}$
          − 3 9.48
          1 4.87

NEXT STEP

Before you go on to the next lesson, I want you to ask yourself a question: "Self, am I getting all of these (or nearly all of these) problems right?" If the answer is yes, then go directly to the next lesson. But if you're having any trouble with the addition or subtraction, then you need to go back and redo Review Lessons 1 and 2 in Section I. Once you've done that, start this chapter over again, and see if you can get everything right.

LESSON | 10

You'll learn how to multiply decimals in this lesson. You'll find out that the big trick is to know where to put the decimal point in your answer. If you can count from 1 to 6, then you can figure out where the decimal goes in your answer.

MULTIPLYING DECIMALS

When you multiply two decimals that are both smaller than 1, your answer, or product, is going to be smaller than either of the numbers you multiplied. Let's prove that by multiplying the two fractions, $\frac{1}{10} \times \frac{1}{10}$. Our answer is $\frac{1}{100}$. Similarly, if we multiply 0.1×0.1, we'll get 0.01, which may be read as one one-hundredth. When you have completed this lesson, you'll be able to do problems like this in your sleep.

The only difference between multiplying decimals and multiplying whole numbers is figuring out where to place the decimal point. For instance, when you multiply 0.5 by 0.5, where do you put the decimal in your answer?

95

You know that $5 \times 5 = 25$. So how much is 0.5×0.5? Is it 0.025, 0.25, 2.5, 25.0, or what? Here's the rule to use: When you multiply two numbers with decimals, add the number of decimal places to the right of the decimal point for both numbers, and then, starting from the right, move the same number of places to find where the decimal point goes in your answer. That probably sounds a lot more complicated than it is.

Let's go back to 0.5×0.5. How many numbers are after the decimal points? There are two numbers after the decimal points: .5 and .5. Now we go to our answer and place the decimal point two places from the right, at 0.25. When you get a few more of these under your belt, you'll be able to do them automatically.

Problem: How much is 0.34×0.63?

Solution:
```
      .34
   × .63
      102
    204
   .2142
```

How many numbers follow the decimals in 0.34 and 0.63? The answer is four. So you start to the right of 2142. and go four places to the left: 0.2142.

Problem: How much is 0.6×0.58?

Solution:
```
      .58
    × .6
    .348
```

How many numbers follow the decimals in 0.6 and 0.58? The answer is three. So you start to the right of 348. and go three places to the left: 0.348.

Here's one that may be a little harder.

Problem: Multiply 50 by 0.72.

Solution:
```
        50
      × .72
       1 00
      35 0
      36.00
```

Again, how many numbers follow the decimal in 0.72? Obviously, two. There aren't any numbers after the decimal point in 50. Starting to the right of 3600. we move two places to the left: 36.00.

This next one is a little tricky. Just follow the rule for placing the decimal point and see if you can get it right.

Problem:
```
      .17
    × .39
```

Solution:
```
      .17
    × .39
     153
     51
     663
```

It looks like I'm stuck. The decimal point needs to go four places to the left. But I've got only three numbers in my answer. So what do I do?

What I need to do is place a zero to the left of 663 and **then** place my decimal point: 0.0663. (I also added the zero that ends up to the left of the decimal point.)

Let's try one more of these.

Problem:
```
      .22
    × .36
```

Solution:
```
       .22
     × .36
      132
       66
      .0792
```

PROBLEM SET

You can easily get the hang of multiplying decimals by working out more problems. So go ahead and do this problem set.

1.	.13 $\times$ .45	**4.**	6.75 $\times$ 9.08
2.	1.4 $\times$ 6.92	**5.**	12.7 $\times$ 6.53
3.	106 $\times$ .57	**6.**	115.81 $\times$ 12.06

Solutions

1.
```
      .13
  ×  .45
      65
      52
    .0585
```

4.
```
      6.75
  ×  9.08
      5400
    60 750
   61.2900
```

2.
```
      1.4
  ×  6.92
      28
    1 26
    8 4
    9.688
```

5.
```
     12.7
  ×  6.53
      381
      6 35
     76 2
    82.931
```

3.
```
      106
  ×  .57
     7 42
    53 0
    60.42
```

6.
```
      115.81
  ×  12.06
      6 9486
    231 620
   1 158 1
  1,396.6686
```

You may have noticed in problem 4 that your answer had a couple of excess zeros, 61.2900. These zeroes can be dropped without changing the value of the answer. So the answer is written as 61.29.

A very common mistake is putting a decimal point in the wrong place. One shortcut to getting the right answer, while avoiding this mistake, is to do a quick approximation of the answer. For example, in problem 4, we're multiplying 6.75 by 9.08. We know that $6 \times 9 = 54$, so we're looking for an answer that's a bit more than 54. Does 6.129 look right to you? How about 612.900? Clearly, the answer 61.2900 looks the best.

NEXT STEP

So far you've added, subtracted, and multiplied decimals. You know what comes next—dividing decimals.

LESSON | 11

In this lesson, you'll learn how to divide decimals. You'll find out that the only difference between dividing decimals and dividing whole numbers is figuring out where to place the decimal point.

DIVIDING DECIMALS

One thing to remember when you're dividing one number by another that's less than 1 is that your answer, or quotient, will be larger than the number divided. For example, if you were to divide 4.0 by 0.5, your quotient would be more than 4.0. We'll come back to this problem in just a minute.

Instead of applying an arithmetic rule as we did when we multiplied decimals, we'll just get rid of the decimals in the divisor (the number by which we divide) and do straight division. I'll work out the first problem to show you just how easy this is.

How much is 4.0 divided by 0.5? Let's do it. Let's set it up as a fraction to start:

$$\frac{4.0}{0.5}$$

Next we'll move the decimal of the numerator one place to the right, and we'll move the decimal of the denominator one place to the right. We can do this because of that good old law of arithmetic that I mentioned earlier: Whatever you do to the top (numerator) you must also do to the bottom (denominator) and vice versa. So we'll multiply the numerator by 10 and the denominator by 10 to get the decimal place moved over one place to the right.

$$\frac{4.0 \times 10}{0.5 \times 10} = \frac{40.}{05.}$$

Then we do simple division:

$$\frac{40}{5} = 8$$

You'll notice that 8 (the answer) is larger than 4 (the number divided). Whenever you divide a number by another number less than 1, your quotient, or answer, will be larger than the number you divided.

How much is 1.59 divided by 0.02? Would you believe that that's the same question as: How much is 159 divided by 2?

The problem can be written this way:

$$\frac{1.59}{.02}$$

Then let's multiply the top and bottom of this fraction by 100. In other words, move the decimal point of the numerator two places to the right, and move the decimal point of the denominator two places to the right:

$$\frac{1.59 \times 100}{.02 \times 100} = \frac{159}{2}$$

We've just reduced the problem to simple division:

$$2 \overline{)159.0} \quad 79.5$$

You'll notice that I added a 0 to 159. By carrying out this division one more decimal place, I avoided leaving a remainder. However, it would have been equally correct to have an answer of 79 with a remainder of 1, or, for that matter, $79\frac{1}{2}$.

Here's one for you to work out.

Problem: How much is 10.62 divided by 0.9?

Solution: $.9 \overline{)10.62}$ =

$$9. \overline{)106.2} = 11.8$$

In this problem, you needed to multiply by 10, so you moved the decimal point **one** place to the right. You multiplied 0.9 by 10 and got 9. Then you multiplied 10.62 by 10 and got 106.2. Then you divided. Very good!

See how you can handle this one.

Problem: Divide 0.4 by 0.25.

Solution:
$$.25 \overline{).4} = 25. \overline{)40.} = 25 \overline{)40} = 5 \overline{)8.0} = 1.6$$

Here's one that may be a bit harder.

Problem: How much is 92 divided by 0.23?

Solution:
$$.23 \overline{)92} = 23. \overline{)9200.} = 23 \overline{)9200} = 400$$
$$\underline{-92XX}$$
$$0$$

PROBLEM SET
Since practice makes perfect in math, let's get in some more practice dividing with decimals. See if you can get all these problems right.

1. $.5 \overline{)10}$ **2.** $.28 \overline{)4.76}$

3. $.06 \overline{)0.9636}$

5. $1.03 \overline{)3.502}$

4. $.42 \overline{)1.3734}$

6. $.88 \overline{)9152}$

Solutions

1.
$$.5 \overline{)10} \qquad = \quad 5 \overline{)\overset{20}{100}}$$

2.
$$.28 \overline{)4.76} \qquad = \quad 28 \overline{)\overset{17}{476}}$$
$$\underline{-28\text{X}}$$
$$196$$
$$\underline{-196}$$

3.
$$.06 \overline{)0.9636} \qquad = \quad 6 \overline{)\overset{16.06}{9\overset{3}{6}.36}}$$

4.
$$.42 \overline{)1.3734} \qquad = \quad 42 \overline{)\overset{3.27}{137.34}}$$
$$\underline{-126 \text{ XX}}$$
$$11\ 3$$
$$\underline{-8\ 4}$$
$$2\ 94$$
$$\underline{2\ 94}$$

5.
$$1.03 \overline{)3.502} \qquad = \quad 103 \overline{)\overset{3.4}{350.2}}$$
$$\underline{-309}$$
$$41\ 2$$
$$\underline{-41\ 2}$$

6.

$$
.88\,\overline{)9152} \quad = \quad 88\,\overline{)915{,}200}
$$

$$
\begin{array}{r}
10{,}400 \\
88\,\overline{)915{,}200} \\
-88\text{X XXX} \\
\hline
35\ 2 \\
-35\ 2 \\
\hline
\end{array}
$$

NEXT STEP

Now that you've added, subtracted, multiplied, and divided decimals, you're ready to work with tenths, hundredths, and thousandths. Indeed, you've already gotten started. In the next lesson you'll go quite a bit further.

LESSON | 12

After you convert fractions
into decimals and then
decimals into fractions,
you'll be ready to add,
subtract, multiply, and
divide tenths, hundredths,
and thousandths.

DECIMALS AND FRACTIONS AS TENTHS, HUNDREDTHS, AND THOUSANDTHS

Decimals can be expressed as fractions, and fractions can be expressed as decimals. For example, one-tenth can be written as a fraction, $\frac{1}{10}$, or as a decimal, .1 (or 0.1). I'll show you how to do these conversions. We'll start out with tenths and hundredths; then we'll move into the thousandths.

TENTHS AND HUNDREDTHS

Can you express the number three-tenths as a fraction and as a decimal? How about forty-five one-hundredths? I'll tell you that forty-five one-hundredths = $\frac{45}{100}$ = 0.45.

How much is three one-hundredths as a fraction and as a decimal? Three one-hundredths = $\frac{3}{100}$ = 0.03. Now see if you can do the problem set on the following page.

PROBLEM SET

Express each of these numbers as a fraction and as a decimal.

1. Seventy-three one-hundredths

2. Nine-tenths

3. Nineteen one-hundredths

4. One one-hundredth

5. Seven-tenths

6. Eleven one-hundredths

Solutions

1. $\frac{73}{100} = 0.73$ **4.** $\frac{1}{100} = 0.01$

2. $\frac{9}{10} = 0.9$ **5.** $\frac{7}{10} = 0.7$

3. $\frac{19}{100} = 0.19$ **6.** $\frac{11}{100} = 0.11$

If you didn't put a zero before the decimal point as shown in the above solutions, were your answers wrong? No. It's customary to put the zero before the decimal point for clarity's sake, but it's not essential to do so.

THOUSANDTHS

Let's move on to thousandths. See if you can write the number three hundred seventeen thousandths as a fraction and as a decimal. Yes, it is $\frac{317}{1000} = 0.317$.

> **Problem:** Write the number forty-one thousandths as a fraction and as a decimal.
>
> **Solution:** $\frac{41}{1000} = 0.041$

Ready for another problem set?

PROBLEM SET

Write each of these numbers as a fraction and as a decimal.

7. Five hundred thirty-two thousandths

8. Nine hundred eighty-four thousandths

9. Sixty-two thousandths

10. Seven thousandths

11. Nine hundred sixty-seven thousandths

12. Two thousandths

Solutions

7. $\frac{532}{1000} = 0.532$

8. $\frac{984}{1000} = 0.984$

9. $\frac{62}{1000} = 0.062$

10. $\frac{7}{1000} = 0.007$

11. $\frac{967}{1000} = 0.967$

12. $\frac{2}{1000} = 0.002$

ADDING AND SUBTRACTING THOUSANDTHS

Now let's add some thousandths.

Problem: See if you can add 1.302 plus 7.951 plus 10.596.

Solution:
$$
\begin{array}{r}
\overset{1}{}\overset{1}{1}.302 \\
7.951 \\
+\ 10.596 \\
\hline
19.849
\end{array}
$$

Problem: Now add 5.002 plus 1.973 plus 4.006 plus 12.758.

Solution:
$$
\begin{array}{r}
\overset{1}{}\overset{1}{}\overset{1}{5}.002 \\
1.973 \\
4.006 \\
+\ \overset{1}{1}2.758 \\
\hline
23.739
\end{array}
$$

Moving right along, here's a subtraction problem for you to work out.

Problem: How much is 10.033 take away 8.975?

Solution:
$$
\begin{array}{r}
{\scriptstyle 9\ 9\ 12\ 1} \\
\cancel{10.033} \\
-\ 8.975 \\
\hline
1.058
\end{array}
$$

Here's one more subtraction problem.

Problem: How much is 14.102 minus 8.479?

Solution:
$$
\begin{array}{r}
{\scriptstyle 3\ 10\ 9\ 1} \\
\cancel{14.102} \\
-\ 8.479 \\
\hline
5.623
\end{array}
$$

How are you doing? Are you ready for another problem set?

PROBLEM SET
Solve each of these problems.

13.	10.071 16.384 + 4.916	**16.**	1.037 − 0.198
14.	15.530 18.107 12.614 + 8.009	**17.**	12.234 − 8.755
15.	23.075 15.928 11.632 + 12.535	**18.**	19.004 − 12.386

Solutions

13.
$$\begin{array}{r} {}^{1\,1}10.{}^{1\,1}071 \\ 16.384 \\ +\ 4.916 \\ \hline 31.371 \end{array}$$

14.
$$\begin{array}{r} {}^{2}15.{}^{1}5{}^{2}30 \\ 18.107 \\ 12.614 \\ +\ 8.009 \\ \hline 54.260 \end{array}$$

15.
$$\begin{array}{r} {}^{1\,2}23.{}^{1\,2}075 \\ 15.928 \\ 11.632 \\ +\ 12.535 \\ \hline 63.170 \end{array}$$

16.
$$\begin{array}{r} {}^{9}\cancel{1}.{}^{12}0\cancel{3}{}^{1}7 \\ -\ 0.198 \\ \hline 0.839 \end{array}$$

17.
$$\begin{array}{r} {}^{1}1{}^{11}2.{}^{12}2\cancel{3}{}^{1}4 \\ -\ 8.755 \\ \hline 3.479 \end{array}$$

18.
$$\begin{array}{r} {}^{8}1{}^{9}\cancel{9}{}^{9}\cancel{0}{}^{1}\cancel{0}4 \\ -\ 12.386 \\ \hline 6.618 \end{array}$$

MULTIPLYING THOUSANDTHS

Multiplying thousandths is really the same as multiplying tenths and hundredths. Let's see if you remember. Work out this problem and be very careful where you place the decimal point.

Problem:
$$\begin{array}{r} 1.375 \\ \times\ 9.084 \end{array}$$

Solution:
$$\begin{array}{r} 1.375 \\ \times\ 9.084 \\ \hline 5500 \\ 11000 \\ 12\ 3750 \\ \hline 12.490\cancel{500} \end{array}$$

The multiplication gives you a product of 12490500. Since there are three numbers after the decimal point of 1.375 and three numbers after the decimal point of 9.084, you need to place the decimal point of your answer six places from the right of 12490500. Moving six places to the left, you get an answer of 12.490500, or 12.4905.

Now do this problem.

Problem:	10.009	**Solution:**	10.009

$$\begin{array}{r} 10.009 \\ \times\ 15.997 \\ \hline 70063 \\ 90081 \\ 90081 \\ 50\ 045 \\ 100\ 09\ \ \ \ \ \\ \hline 160.113973 \end{array}$$

Again, you need to move your decimal point six places to the left of your product. Start at the extreme right and count six places to the left, which gives you an answer of 160.113973.

Now I'd like you to do the following problem set.

PROBLEM SET

Here are some problems for you to practice.

19. $\begin{array}{r} 4.350 \\ \times\ 1.281 \\ \hline \end{array}$ **22.** $\begin{array}{r} 3.692 \\ \times\ 8.417 \\ \hline \end{array}$

20. $\begin{array}{r} 5.728 \\ \times\ 2.043 \\ \hline \end{array}$ **23.** $\begin{array}{r} 16.559 \\ \times\ 12.071 \\ \hline \end{array}$

21. $\begin{array}{r} 10.539 \\ \times\ 4.167 \\ \hline \end{array}$ **24.** $\begin{array}{r} 21.006 \\ \times\ 36.948 \\ \hline \end{array}$

Solutions

19.
```
        4.350
     × 1.281
        4350
      34800
       8700
      4 350
    5.572350
```

22.
```
        3.692
     × 8.417
       25844
        3692
      1 4768
     29 536
   31.075564
```

20.
```
        5.728
     × 2.043
       17184
       22912
     11 4560
   11.702304
```

23.
```
       16.559
    × 12.071
       16559
     1 15913
     33 1180
     165 59
  199.883689
```

21.
```
       10.539
     × 4.167
       73773
       63234
      1 0539
     42 156
   43.916013
```

24.
```
       21.006
    × 36.948
      168048
       84024
      18 9054
     126 036
     630 18
   776.129688
```

I'd like you to take another look at the problem set you just did. It's very easy to put the decimal point in the wrong place in your answer, so I'd like to give you a helpful hint. That hint is to estimate your answer before you even do the multiplication. In problem 19, you would expect an answer that's somewhat more than 4 because you're multiplying a number somewhat larger than 4 by a number a bit larger than 1. So if you ended up with 55.72350 or 0.5572350, you can see that neither of those answers make sense.

In problem 20, you would estimate your answer to be somewhat larger than 10 (since 5 × 2 = 10). We ended up with 11.702304, which certainly looks right. Go ahead and carry out this reality check on the answers to problems 21 through 24. And remember that when you're multiplying decimals, it really pays to estimate your answer before you even do the problem.

DIVIDING THOUSANDTHS

So far you've added, subtracted, and multiplied thousandths. You probably know what comes next. Division! Just remember to get rid of the decimals in the divisor, and the problem becomes straightforward division.

Problem: Divide 2.112 by 0.132.

Solution:

$$.132 \overline{)2.112} \quad = \quad 132, \overline{)2112.} \quad = \quad 132 \overline{)2112} \begin{array}{r} 16 \\ \end{array}$$

$$\begin{array}{r} -132X \\ \hline 792 \\ -792 \end{array}$$

Here's one that's slightly more difficult.

Problem: Divide 0.0645 by 0.043.

Solution:

$$.043 \overline{).0645} \quad = \quad 043, \overline{)064.5} \quad = \quad 43 \overline{)64.5} \begin{array}{r} 1.5 \\ \end{array}$$

$$\begin{array}{r} -43 X \\ \hline 21\ 5 \\ -21\ 5 \end{array}$$

PROBLEM SET

Here's your chance to do several division problems.

25. Divide 17.85 by 0.525.

26. Divide 2.0912 by 1.307.

27. Divide 1.334 by 0.046.

28. Divide 138.4 by 0.008.

29. Divide 7.2054 by 4.003.

30. Divide .26588 by 1.156.

Solutions

25.

$$.525 \overline{)17.85} \quad = \quad 525 \overline{)17850} \atop \begin{array}{r} 34 \\ \hline 17850 \\ -\,1575 \\ \hline 2100 \\ -\,2100 \end{array}$$

26.

$$1.307 \overline{)2.0912} \quad = \quad 1307 \begin{array}{r} 1.6 \\ \overline{)2091.2} \\ -\,1307\,X \\ \hline 784\,2 \\ -\,784\,2 \end{array}$$

27.

$$.046 \overline{)1.334} \quad = \quad 46 \begin{array}{r} 29 \\ \overline{)1334} \\ -\,92X \\ \hline 414 \\ -\,414 \end{array}$$

28.

$$.008 \overline{)138.4} \quad = \quad 8 \begin{array}{r} 17,300 \\ \overline{)13\,{}^{5}8\,{}^{2}400} \end{array}$$

29.

$$4.003 \overline{)7.2054} \quad = \quad 4003 \begin{array}{r} 1.8 \\ \overline{)7205.4} \\ -\,4003\,X \\ \hline 3202\,4 \\ -\,3202\,4 \end{array}$$

30.

$$1.156 \overline{).26588} \quad = \quad 1156 \begin{array}{r} .23 \\ \overline{)265.88} \\ -\,231\,2X \\ \hline 34\,68 \\ -\,34\,68 \end{array}$$

One day, when you fill up at a gas station, your car's odometer reads 28,106.3. The next time you fill up, your odometer reads 28,487.1. If you just bought 14.2 gallons of gas, how many miles per gallon did you get, rounded to the tenths place?

Solution

$$
\begin{array}{r}
28{,}487.1 \\
-\ 28{,}106.3 \\
\hline
380.8
\end{array}
$$

$$
\frac{380.8 \text{ miles}}{14.2 \text{ gallons}} \quad = \quad 142
$$

$$
\begin{array}{r}
26.8 \ \text{miles per gallon} \\
142\ )\overline{3808.0} \\
\underline{284} \\
968 \\
\underline{-852} \\
1160 \\
\underline{-1136}
\end{array}
$$

NEXT STEP

How are you doing? If you're getting everything—or almost everything—right, then go directly to Lesson 13. But if you're having any trouble at all, then you'll need to review some of the material you've already covered. For instance, if you're having trouble adding or subtracting decimals, you'll need to review Lesson 9 as well as the second part of this lesson. If you're not doing well multiplying decimals, then you'll need to rework your way through Lesson 10 and the third part of this lesson. And if you're at all shaky on dividing decimals, then you'll need to review Lesson 11 and the last part of this lesson before moving on to the next lesson.

LESSON | 13

When you've finished this lesson, you'll be able to convert a decimal into a fraction, which involves getting rid of the decimal point. You've already done some conversion of fractions into decimals. When you convert a fraction into a decimal, you're dividing the denominator into the numerator and adding a decimal point.

CONVERTING FRACTIONS INTO DECIMALS AND DECIMALS INTO FRACTIONS

In the last lesson you expressed tenths and hundredths as fractions and as decimals. Tenths and hundredths are easy to work with, but some other numbers are not as simple. You'll learn to do more difficult conversions in this lesson. Let's start by converting fractions into decimals. Then we'll move into expressing decimals as fractions.

FRACTIONS TO DECIMALS

How would you convert $\frac{17}{20}$ into a decimal? There are actually two ways. Remember the arithmetic law that says what we do to the top (numerator) of a fraction, we must also do to the bottom (denominator)? That's one way to do it. Go ahead and convert $\frac{17}{20}$ into hundredths.

$$\frac{17 \times 5}{20 \times 5} = \frac{85}{100} = 0.85$$

Now let's use the second method to convert the fraction $\frac{17}{20}$ into a decimal. Are you ready? Every fraction can be converted into a decimal by dividing its denominator (bottom) into its numerator (top). Go ahead and divide 20 into 17.

$$
\begin{array}{r}
.85 \\
20 \overline{)17.00} \\
-16\ 0\cancel{X} \\
1\ 00 \\
-1\ 00 \\
\end{array}
$$

Problem: Use both methods to convert $\frac{19}{50}$ into a decimal.

Solution: $\frac{19 \times 2}{50 \times 2} = \frac{38}{100} = 0.38$ or

$$
\begin{array}{r}
.38 \\
50 \overline{)19.00} \\
-15\ 0\cancel{X} \\
4\ 00 \\
-4\ 00 \\
\end{array}
$$

We've done two problems so far where we could convert the denominator to 100. But we're not always that lucky. See if you can convert the following fraction into a decimal.

Problem: Convert $\frac{3}{8}$ into a decimal.

Solution:
$$\frac{3}{8} = 8 \overline{)3.0\overset{6}{0}\overset{4}{0}} \quad .375$$

Sometimes we have fractions that can be reduced before being converted into decimals. See what you can do with the next one.

Problem: Convert $\frac{9}{12}$ into a decimal.

Solution:

$$\frac{9}{12} = \frac{3}{4} \qquad 4\overline{)3.0\overset{2}{0}} \quad \begin{matrix} .75 \end{matrix}$$

It often pays to reduce a fraction to its lowest possible terms because that will simplify the division. It's easier to divide 4 into 3 than to divide 12 into 9. By the way, when I divided 4 into 3, I placed a decimal point after the 3 and then added a couple of zeroes. The number 3 may be written as 3.0, and we may add as many zeroes after the decimal point as we wish.

Now let's see if you can handle this problem set.

PROBLEM SET
Convert each of these fractions into a decimal.

1. $\frac{3}{5} =$ **4.** $\frac{5}{8} =$

2. $\frac{13}{25} =$ **5.** $\frac{6}{15} =$

3. $\frac{44}{200} =$ **6.** $\frac{69}{300} =$

Solutions

1. $\frac{3}{5} = \frac{3 \times 20}{5 \times 20} = \frac{60}{100} = 0.6$ **4.** 0.625

$$\frac{5}{8} = 8\overline{)5.0\overset{2\ 4}{0}0}$$

2. $\frac{13}{25} = \frac{13 \times 4}{25 \times 4} = \frac{52}{100} = 0.52$ **5.** $\frac{6}{15} = \frac{2}{5} = \frac{2 \times 20}{5 \times 20} = \frac{40}{100} = 0.4$

3. $\frac{44}{200} = \frac{22}{100} = 0.22$ **6.** $\frac{69}{300} = \frac{23}{100} = 0.23$

DECIMALS TO FRACTIONS
You're going to catch a break here. Decimals can be converted into fractions in two easy steps. If it were a dance, we'd call it the easy two-step. First I'll do one. I'm going to convert the decimal 0.39 into a fraction. All I have to do is get rid of the decimal point by moving it two places to the right, and then placing the 39 over 100: $0.39 = \frac{39}{100}$.

Here's a couple for you to do.

Problem: Convert 0.73 into a fraction.

Solution: $0.73 = \frac{73}{100}$

Problem: Now convert 0.4 into a fraction.

Solution: $0.4 = \frac{4}{10}$

Since we like to convert fractions into their lowest terms, let's change $\frac{4}{10}$ into $\frac{2}{5}$. For tenths and hundredths, however, you don't necessarily have to do this. It's the mathematical equivalent of crossing your *t*'s and dotting your *i*'s. The fraction $\frac{4}{10}$ *is* mathematically correct, but there are some people out there who will insist that every fraction be reduced to its lowest terms. Luckily for you, I am not one of them, at least when it comes to tenths and hundredths.

Are you ready for a problem set? Good, because here comes one now.

PROBLEM SET
Convert each of these decimals into fractions.

7. $0.5 =$ **10.** $0.97 =$

8. $0.65 =$ **11.** $0.09 =$

9. $0.18 =$ **12.** $0.1 =$

Solutions

7. $0.5 = \frac{5}{10} = \frac{1}{2}$ **10.** $0.97 = \frac{97}{100}$

8. $0.65 = \frac{65}{100} = \frac{13}{20}$ **11.** $0.09 = \frac{9}{100}$

9. $0.18 = \frac{18}{100} = \frac{9}{50}$ **12.** $0.1 = \frac{1}{10}$

CONVERTING THOUSANDTHS

Let's wrap this up by converting some decimal thousandths into fractional thousandths, and then some fractional thousandths into decimal thousandths.

Problem: Convert $\frac{247}{1000}$ into a decimal.

Solution: $\frac{247}{1000} = 0.247$

Problem: Convert $\frac{19}{1000}$ into a decimal.

Solution: $\frac{19}{1000} = 0.019$

Problem: Let's shift gears and change 0.804 into a fraction.

Solution: $0.804 = \frac{804}{1000} \left(= \frac{201}{250}\right)$

One more, then we'll do a problem set.

Problem: Convert 0.003 into a fraction.

Solution: $0.003 = \frac{3}{1000}$

PROBLEM SET

Convert these fractions into decimals.

13. $\frac{815}{1000} =$

14. $\frac{43}{1000} =$

15. $\frac{5}{1000} =$

Convert these decimals into fractions.

16. $0.153 =$

17. $0.001 =$

18. $0.089 =$

Solutions

13. $\frac{815}{1000} = 0.815$ **16.** $0.153 = \frac{153}{1000}$

14. $\frac{43}{1000} = 0.043$ **17.** $0.001 = \frac{1}{1000}$

15. $\frac{5}{1000} = 0.005$ **18.** $0.089 = \frac{89}{1000}$

Do you remember your profitable transaction with Microsoft stock? You bought 100 shares at $109\frac{3}{4}$ and sold them at $116\frac{3}{8}$. Let's calculate your profit, this time using decimals.

Solution:

You paid $198.75 for each share, which you sold at $116.375.

$116.375
$\underline{-109.750}$
$6.625

So you made a profit of $6.625 on each of 100 shares, or a total of $662.50.

NEXT STEP

Now that you can convert fractions into decimals and decimals into fractions, you're ready to do some fast multiplication and division. Actually, in the next lesson, all you'll need to do is move around some decimal points and add or subtract some zeros.

LESSON | 14

Doing fast multiplication and division can be a whole lot of fun. When you've completed this lesson, you'll be able to multiply a number by 1,000 in a fraction of a second and divide a number by 1,000 just as quickly.

FAST MULTIPLICATION AND FAST DIVISION

Wouldn't it be great to know some math tricks, so you could amaze your friends with a speedy answer to certain math questions? There are shortcuts you can use when multiplying or dividing by tens, hundreds, or thousands. Let's start with some multiplication problem shortcuts.

MULTIPLYING WHOLE NUMBERS BY 10, 100, AND 1,000

Try to answer the next question as quickly as possible before you look at the solution.

Problem: Quick, how much is 150×100?

Solution: The answer is 15,000. What I did was add two zeros to 150.

Problem: How much is $32 \times 1,000$?

Solution: I'll bet you knew it was 32,000.

So one way of doing fast multiplication is by adding zeros. Before we talk about the other way of doing fast multiplication, I'd like you to do this problem set.

PROBLEM SET

Multiply each of these numbers by 10.

1. 410

2. 1

3. 1,000

Multiply each of these numbers by 100.

4. 50

5. 629

6. 3,000

Multiply each of these numbers by 1,000.

7. 1,000

8. 40

9. 232

Solutions

1. $410 \times 10 = 4,100$

2. $1 \times 10 = 10$

3. $1,000 \times 10 = 10,000$

4. $50 \times 100 = 5,000$

5. $629 \times 100 = 62,900$

6. $3,000 \times 100 = 300,000$

7. $1,000 \times 1,000 = 1,000,000$

8. $40 \times 1,000 = 40,000$

9. $232 \times 1,000 = 232,000$

MULTIPLYING DECIMALS BY 10, 100, AND 1,000

Multiplying decimals by 10, 100, and 1,000 is different from multiplying whole numbers by them because you can't just add zeros. Take a stab at the following questions.

Problem: How much is 1.8×10?

Solution: You can't add a zero to 1.8, because that would leave you with 1.80, which has the same value as 1.8. But what you could do is move the decimal point one place to the right, 18., which gives you 18.

Problem: How much is 10.67×100?

Solution: Just move the decimal point two places to the right, 1067. and you get 1,067.

Now here's one that's a little tricky.

Problem: How much is 4.6×100?

Solution: First we add a zero to 4.6, making it 4.60. We can add as many zeros as we want after a decimal, because that won't change its value. Once we've added the zero, we can move the decimal point two places to the right: 460.

By convention, we don't use decimal points after whole numbers like 460, so we can drop the decimal point.

Problem: How much is 9.2×100?

Solution: $9.2 \times 100 = 9.20 \times 100 = 920$

Problem: How much is $1.573 \times 1,000$?

Solution: The answer is 1,573. All you needed to do was move the decimal point three places to the right.

To summarize: When you multiply a decimal by 10, move the decimal point one place to the right. When you multiply a decimal by 100, move the decimal point two places to the right. When you multiply a decimal by 1,000, move the decimal point three places to the right.

Problem: Now multiply $10.4 \times 1,000$.

Solution: $10.4 \times 1,000 = 10.400 \times 1,000 = 10,400$

PROBLEM SET
Multiply these numbers by 10.

10. 6.2

11. 1.4

12. 30.22

Multiply these numbers by 100.

13. 11.2

14. 1.44

15. 50.3

Multiply these numbers by 1,000.

16. 14.02

17. 870.9

18. 12.91

Solutions

10. $6.2 \times 10 = 62$

11. $1.4 \times 10 = 14$

12. $30.22 \times 10 = 302.2$

13. $11.2 \times 100 = 11.20 \times 100 = 1{,}120$

14. $1.44 \times 100 = 144$

15. $50.3 \times 100 = 50.30 \times 100 = 5{,}030$

16. $14.02 \times 1{,}000 = 14.020 \times 1{,}000 = 14{,}020$

17. $870.9 \times 1{,}000 = 870.900 \times 1{,}000 = 870{,}900$

18. $12.91 \times 1{,}000 = 12.910 \times 1{,}000 = 12{,}910$

FAST DIVISION

Fast division is the reverse of fast multiplication. Instead of adding zeros, you take them away. And instead of moving the decimal point to the right, you move it to the left.

DIVIDING BY TEN

Start off by taking zeros away from the first number in the next question.

Problem: How much is 140 divided by 10?

Solution: The answer is 14. All you did was get rid of the zero.

Problem: How much is 1,300 divided by 10?

Solution: The answer is 130.

So far, so good. But what do you do if there are no zeros to get rid of? Then you must move the decimal point one place to the left.

Problem: For instance, how much is 263 divided by 10?

Solution: $263 \div 10 = 26.30 = 26.3$

Problem: How much is 1,094 divided by 10?

Solution: $1,094 \div 10 = 109.40 = 109.4$

PROBLEM SET

Divide each of these numbers by 10.

19. 10

20. 4,590

21. 383

22. 1,966

23. 1.77

24. 68.2

Solutions

19. $10 \div 10 = 1$

20. $4,590 \div 10 = 459$

21. $383 \div 10 = 38.3$

22. $1,966 \div 10 = 196.6$

23. $1.77 \div 10 = 0.177$

24. $68.2 \div 10 = 6.82$

DIVIDING BY 100

Now we'll move on to dividing by 100.

Problem: How much is 38.9 divided by 100?

Solution: The answer is 0.389—all you need to do is move the decimal point two places to the left.

Problem: How much is 0.4 divided by 100?

Solution: $0.4 \div 100 = 00.4 \div 100 = 0.004$. You can place as many zeros as you wish to the left of a decimal without changing its value. So $00.4 = 0.4$.

Problem: How much is $0.06 \div 100$?

Solution: $0.06 \div 100 = 00.06 \div 100 = 0.0006$

Problem: How much is 4 divided by 100?

Solution: $4 \div 100 = 04.0 \div 100 = 0.04$

You probably remember that you can add zeros after a decimal point without changing its value. I placed zeros to the left of 4 because if you move the decimal point to the left of a whole number, it's understood that you'll need to add zeros.

Problem: How much is $56 \div 100$?

Solution: $56 \div 100 = 56.0 \div 100 = 0.56$

PROBLEM SET

Divide each of these numbers by 100.

25. 89.6

26. 239

27. 1.4

28. 16

29. 2

30. 0.9

Solutions

25. $89.6 \div 100 = 0.896$

28. $16 \div 100 = 0.16$

26. $239 \div 100 = 2.39$

29. $2 \div 100 = 0.02$

27. $1.4 \div 100 = 0.014$

30. $0.9 \div 100 = 0.009$

DIVIDING BY 1,000

Here are some problems for you to practice dividing by 1,000.

Problem: How much is 6,072.5 divided by 1,000?

Solution: The answer is 6.0725. All you needed to do was move the decimal point three places to the left.

Problem: How much is 400,000 divided by 1,000?

Solution: The answer is 400. All you needed to do here was to drop three zeros.

Problem: How much is 752 divided by 1,000?

Solution: $752 \div 1,000 = 0.752$

Problem: How much is $39 \div 1,000$?

Solution: $39 \div 1,000 = 0.039$

Problem: How much is 0.2 divided by 1,000?

Solution: Just move the decimal point three places to the left: 0.0002.

PROBLEM SET

Divide each of these numbers by 1,000.

31. 309.6

34. 150

32. 4.8

35. 3

33. 60,000

36. 0.5

Solutions

31. $309.6 \div 1,000 = 0.3096$

34. $150 \div 1,000 = 0.15$

32. $4.8 \div 1,000 = 0.0048$

35. $3 \div 1,000 = 0.003$

33. $60,000 \div 1,000 = 60$

36. $0.5 \div 1,000 = 0.0005$

SUMMARY

So far you've multiplied and divided by 10, 100, and 1,000. Let's summarize the procedures you've followed.

- To multiply by 10, you add a zero or move the decimal point one place to the right.
- To divide by 10, you drop a zero or move the decimal point one place to the left.
- To multiply by 100, you add two zeros or move the decimal point two places to the right.
- To divide by 100, you drop two zeros or move the decimal point two places to the left.
- To multiply by 1,000, you add three zeros or move the decimal point three places to the right.
- To divide by 1,000, you drop three zeros or move the decimal point three places to the left.

Don't worry, you won't have to memorize all these rules. All you'll need to do when you want to multiply a number by 10, 100, or 1,000 is ask yourself how you can make this number larger. Do you do it by tacking on zeros or by moving the decimal point to the right? As you get used to working with numbers, doing this will become virtually automatic.

Similarly, when you divide, just ask yourself how you can make this number smaller. Do you do it by dropping zeros, or by moving the decimal point to the left? Again, with experience you'll be doing these problems instinctively.

Another way of expressing a division problem is to ask: How much is one-tenth of 50? Or how much is one one-hundredth of 7,000? One-tenth of 50 obviously means how much is 50 divided by 10, so the answer is 5. And one one-hundredth of 7,000 means how much is 7,000 divided by 100, which is 70.

Problem: How much is one-tenth of 16,000?

Solution: The answer is 1,600.

Problem: How much is one-tenth of 1.3?

Solution: The answer is 0.13.

Problem: How much is one one-hundredth of 9?

Solution: The answer is 0.09.

Problem: And how much is one one-thousandth of 8.6?

Solution: The answer is 0.0086.

PROBLEM SET
Find one-tenth of each of these numbers.

37. 800

40. 1.9

38. 2

41. 0.03

39. 43

42. 0.2

Find one one-hundredth of each of these numbers.

43. 500,000   **46.** 3.6

44. 89.3   **47.** 400

45. 57   **48.** 0.1

Find one one-thousandth of each of these numbers.

49. 750   **52.** 14.2

50. 0.9   **53.** 116

51. 6,000   **54.** 600

Solutions

37. $800 \times 0.1 = 80$

38. $2 \times 0.1 = 0.2$

39. $43 \times 0.1 = 4.3$

40. $1.9 \times 0.1 = 0.19$

41. $0.03 \times 0.1 = 0.003$

42. $0.2 \times 0.1 = 0.02$

43. $500,000 \times 0.01 = 5,000$

44. $89.3 \times 0.01 = 0.893$

45. $57 \times 0.01 = 0.57$

46. $3.6 \times 0.01 = 0.036$

47. $400 \times 0.01 = 4$

48. $0.1 \times 0.01 = 0.001$

49. $750 \times 0.001 = 0.75$

50. $0.9 \times 0.001 = 0.0009$

51. $6{,}000 \times 0.001 = 6$

52. $14.2 \times 0.001 = 0.0142$

53. $116 \times 0.001 = 0.116$

54. $600 \times 0.001 = 0.6$

NEXT STEP

I told you this lesson would be a whole lot of fun. In the next lesson, you'll get a chance to apply everything you've learned about decimals in this entire section.

LESSON | 15

After introducing coins as decimals of a dollar, this lesson will help you apply what you've learned about decimals to real-world problems that involve addition, subtraction, multiplication, and division.

APPLICATIONS

Some of the applications of the math you've done in this section are money problems. So before you actually do any problems, let's talk a little about the U.S. dollar. The dollar can be divided into fractions or decimals. There are 100 cents in a dollar. If a dollar is 1, or 1.0, then how much is a half dollar (50 cents) as a fraction of a dollar and as a decimal of a dollar? It's $\frac{1}{2}$ or 0.5 (or 0.50).

Problem: Write each of these coins as a fraction of a dollar and as a decimal of a dollar:

a. A penny

b. A nickel

   c.   A dime

   d.   A quarter

Solution:

   a.   A penny $= \frac{1}{100} = 0.01$

   b.   A nickel $= \frac{5}{100}$ (or $\frac{1}{20}$) $= 0.05$

   c.   A dime $= \frac{10}{100}$ (or $\frac{1}{10}$) $= 0.1$ (or 0.10)

   d.   A quarter $= \frac{25}{100}$ (or $\frac{1}{4}$) $= 0.25$

Before you begin the problem set, let me say a few words about rounding your answers. Suppose your answer came to $14.9743. Rounded to the nearest penny, your answer would be $14.97. If your answer were $30.6471, rounded to the nearest penny it would come to $30.65. So whenever this applies, round your answers to the nearest penny.

PROBLEM SET

Do all of these problems, and then check your work with the solutions that follow.

1.   If you had a half dollar, three quarters, eight dimes, six nickels, and nine pennies, how much money would you have all together?

2.   If your weekly salary is $415.00, how much do you take home each week after deductions are made for federal income tax ($82.13), state income tax ($9.74), Social Security and Medicare ($31.75), and retirement ($41.50)?

3.   You began the month with a checking account balance of $897.03. During the month you wrote checks for $175.00, $431.98, and $238.73, and you made deposits of $300.00 and $286.17. How much was your balance at the end of the month?

4.   Carpeting costs $7.99 a yard. If Jose buys 12.4 yards, how much will it cost him?

5.   If cashews cost $6.59 a pound, how much would it cost to buy two and a quarter pounds?

6. Sheldon Chen's scores in the diving competition were 7.2, 6.975, 8.0, and 6.96. What was his total score?

7. If gasoline cost $1.399 a gallon, how much would it cost to fill up a tank that had a capacity of $14\frac{3}{4}$ gallons?

8. The winners of the World Series received $958,394.31. If this money was split into 29.875 shares, how much would one share be worth?

9. If you bought $3\frac{1}{2}$ pounds of walnuts at $4.99 a pound and $1\frac{1}{4}$ pounds of peanuts at $2.39 a pound, how much would you spend all together?

10. The Coney Island Historical Society had sales of $3,017.93. After paying $325 in rent, $212.35 in advertising, $163.96 in insurance, and $1,831.74 in salaries, how much money was left in profits?

11. If gold cost $453.122 an ounce, how much would $\frac{3}{8}$ of an ounce cost?

12. Jessica owned 1.435 shares, Karen owned 2.008 shares, Jason owned 1.973 shares, and Elizabeth owned 2.081 shares. How many shares did they own in total?

13. Wei Wong scored 9.007 in gymnastics. Carlos Candellario scored 8.949. How much higher was Wei Wong's score?

14. On Tuesday Bill drove 8.72 hours, averaging 53.88 miles per hour. On Wednesday he drove 9.14 hours, averaging 50.91 miles per hour. How many miles did he drive on Tuesday and Wednesday?

15. One meter is equal to 39.37 inches. How many inches are there in 70.26 meters?

16. Michael studied for 17.5 hours over a period of 4.5 days. On average, how much did he study each day?

17. All the people working at the Happy Valley Industrial Park pooled their lottery tickets. When they won $10,000,000, each got a 0.002 part share. How much money did each person receive?

18. Daphne Dazzle received 2.3 cents for every ticket sold to her movie. If 1,515,296 tickets were sold, how much money did she receive?

19. A cheese store charged $3.99 a pound for American cheese, $3.49 a pound for Swiss cheese, and $4.99 a pound for brie. If it sold 10.4 pounds of American, 16.3 pounds of Swiss, and 8.7 pounds of brie, how much were its total sales?

20. A prize of $10,000,000 is awarded to three sisters. Eleni receives one-tenth, Justine receives one-tenth, and Sophie receives the rest. How much are their respective shares?

21. Elizabeth and Daniel received cash bonuses equal to one one-hundredth of their credit card billings. If Elizabeth had a billing of $6,790.22 and Daniel had a billing of $5,014.37, how much cash bonus did each of them receive?

22. Con Edison charges 4.3 cents per kilowatt hour. How much does it charge for 1,000 kilowatt hours?

Solutions

1.
$.50
.75
.80
.30
+ .09
$2.44

2.
$82.13
9.74
31.75
+ 41.50
$165.12

$ 415.00
− 165.12
$249.88

3.

$897.03	$175.$\overset{1}{0}$0
300.00	431.98
+ 286.17	+ 238.73

$1,4$\overset{7\ 12\ 11\ 1}{\cancel{8}\cancel{3}.\cancel{2}}$0 $845.71

− 845.71

$637.49

4.

$7.99
× 12.4
‾‾‾‾‾‾
3 196
15 98
79 9
‾‾‾‾‾‾
$99.076   or $99.08

5.

$2\frac{1}{4} = 2.25$

$6.59
× 2.25
‾‾‾‾‾‾
3295
1 318
13 18
‾‾‾‾‾‾
$14.8275   or $14.83

6.

7.200
6.975
8.000
6.960
‾‾‾‾‾‾
29.135

7.

$14\frac{3}{4} = 14.75$

$1.399
× 14.75
‾‾‾‾‾‾
6995
9793
5 596
13 99
‾‾‾‾‾‾
$20.63525   or $20.64

8.
$$29.875 \overline{)\ \$958{,}394.31}$$

$$
\begin{array}{r}
32\ 080.14 \\
\end{array}
$$
=29,875)$958,394,310.00

896 25

62 144

59 750

2 394 31

2 390 00

4 310 0

2 987 5

1 322 50

1 195 00

127 50

= $32,080.14

9.

$3\frac{1}{2} = 3.5$ $1\frac{1}{4} = 1.25$

$4.99 $2.39

× 3.5   × 1.25

2 495   11 95

14 97   47 8

$17.465   or $17.47   2 39

                                    $2.9875   or $2.99

$17.465   17.47

+ 2.9875   + 2.99

$20.4525   or $20.45 $20.46

10. $325.00 $ 3,017.93

            212.35   − 2,533.05

            163.96 $ 484.88

     + 1,831.74

     $ 2,533.05

11. $\frac{3}{8} = 0.375$

$453.122
×   .375
2 265610
31 71854
135 9366
169.920750   or $169.92

12.   1.435
2.008
1.973
+ 2.081
7.497

13.   ⁸ ⁹⁹¹
9.007
− 8.949
0.058

14.   53.88   50.91
× 8.72   × 9.14
1 0776   2 0364
37 716   5 091
431 04   458 19
469.8336   465.3174

469.8336
+ 465.3174
935.1510

15.   39.37
× 70.26
2 3622
7 874
2 755 90
2,766.1362

16.

$$4.5 \overline{)17.5} \quad = \quad 45 \overline{)175} \quad = \quad 9 \overline{)35.000} \quad \text{or } 3.89$$

$$
\begin{array}{r}
3.888 \\
9 \overline{)35.000} \\
\underline{27} \\
8\,0 \\
\underline{7\,2} \\
80 \\
\underline{72} \\
80 \\
\underline{72} \\
8
\end{array}
$$

17.

$$
\begin{array}{r}
10{,}000{,}000 \\
\times \qquad .002 \\
\hline
\$\,20{,}000.000
\end{array}
$$

18.

$$
\begin{array}{r}
1{,}515{,}296 \\
\times \qquad \$.023 \\
\hline
4\;545\;888 \\
30\;305\;92 \\
\hline
\$34{,}851.808 \quad \text{or } \$34{,}851.81
\end{array}
$$

19.

$$
\begin{array}{r}
\$3.99 \\
\times 10.4 \\
\hline
1\;596 \\
39\;90 \\
\hline
\$41.496
\end{array}
\qquad
\begin{array}{r}
\$3.49 \\
\times 16.3 \\
\hline
1\;047 \\
20\;94 \\
34\;9 \\
\hline
\$56.887
\end{array}
\qquad
\begin{array}{r}
\$4.99 \\
\times 8.7 \\
\hline
3\;493 \\
39\;92 \\
\hline
\$43.413
\end{array}
$$

$$
\begin{array}{r}
\$41.496 \\
56.887 \\
+\;43.413 \\
\hline
\$141.796 \quad \text{or } \$141.80
\end{array}
$$

20. Eleni received $1,000,000; Justine received $1,000,000; Sophie received $8,000,000.

21. Elizabeth received $67.90 and Daniel received $50.14.

22. $0.043 \times 1,000 = \$4,300$

Suppose you wanted to compare your cost per mile using Mobil regular gasoline, which costs $1.199 per gallon and Mobil premium, which costs $1.399 per gallon. If Mobil regular gives you 26.4 miles per gallon (highway driving) and Mobil premium gives you 31.7 miles per gallon, which gas gives you the lower cost per mile? Hint: How much does it cost to drive one mile using both gases?

Solution:

regular: $\dfrac{\$1.199}{26.4} = 4.54$ cents/mile

premium: $\dfrac{\$1.399}{31.7} = 4.41$ cents/mile

Almost everyone who plays the lottery knows that they are overpaying for a ticket. They have only an infinitesimal chance of winning, but, as the tag line of an ad touting the New York State lottery says, "Hey, you never know." OK, let's assume a payoff of $15 million. If any $2-ticketholder's chance of winning were one in 20 million, how much is that ticket really worth?

Solution:

$$\frac{\$15\ \text{million}}{20\ \text{million}} = \frac{\$15}{20} = \frac{\$3}{4} = \$.75\ (75\ \text{cents})$$

Since the millions cancel out, why bother to write out all the zeros?

NEXT STEP

Okay, three sections down, one to go. Once again, let me ask you how things are going. If they're going well, then you're ready for the final section, which introduces percentages. If not, you know the drill. Go back over anything that needs going over. Just let your conscience be your guide.

SECTION IV

PERCENTAGES

Percentages are the mathematical equivalent of fractions and decimals. For example, $\frac{1}{2} = 0.5 = 50\%$. In baseball, a 300 batter is someone who averages three hundred base hits every thousand times at bat, which is the same as thirty out of a hundred ($\frac{30}{100}$ or 30%) or three out of ten ($\frac{3}{10}$). It means he gets a hit 30% of the time that he comes to bat.

Let's take a close look at the relationship among decimals, fractions, and percentages. We'll begin with the fraction, $\frac{1}{100}$. How much is $\frac{1}{100}$ as a percent? It's 1%. And how much is the decimal, 0.01, as a percent? Also 1%.

That means, then, that $\frac{1}{100} = 0.01 = 1\%$. How about 0.10 and $\frac{10}{100}$? As a percent, they're both equal to 10%.

Now I'm going to throw you a curve ball. How much is the number 1 as a decimal, a fraction, and as a percent? The number *one* may be written as 1.0, $\frac{1}{1}$ (or $\frac{100}{100}$), or as 100%.

It's easy to go from fractions and decimals to percents if you follow the procedures outlined in this section. It doesn't matter that much whether you can verbalize these procedures. In math the bottom line is always the same—coming up with the right answer.

When you have completed this section, you will be able to find percentages, convert percentages into fractions and decimals, and find percentage changes, percentage distribution, and percentages of numbers. In short, you will have learned everything you will ever need to know about percentages.

LESSON | 16

In this lesson, you'll learn how to convert decimals into percents and percents into decimals. You'll find out how and when to move the decimal point for each type of conversion. This easy conversion process will lead you to the more difficult process of converting between fractions and percents in the next lesson.

CONVERTING DECIMALS INTO PERCENTS AND PERCENTS INTO DECIMALS

Decimals can be converted into percents by moving their decimal points two places to the right and adding a percent sign. Conversely, percents can be converted into decimals by removing the percent sign and moving their decimal points two places to the left.

CONVERTING DECIMALS TO PERCENTS

You know that the same number can be expressed as a fraction, as a decimal, or as a percent. For example, $\frac{1}{4} = 0.25$. Now what percent is $\frac{1}{4}$ and 0.25 equal to?

The answer is 25%. Just think of these numbers as money: one quarter equals 25 cents, or $0.25, or 25% of a dollar.

Here's how to figure it out. Start with the decimal, 0.25. Let's convert it into a percent. What you do is move the decimal point two places to the right and add a percent sign:

.25 = 25.% = 25%

When we have a whole number like 25, we don't bother with the decimal point. If we wanted to, we could, of course, write 25% like this: 25.0%.

PROBLEM SET

I'd like you to convert a few decimals into percents.

1. 0.32 =

4. 0.03 =

2. 0.835 =

5. 0.41 =

3. 1.29 =

Solutions

1. 0.32 = 32%

4. 0.03 = 3%

2. 0.835 = 83.5%

5. 0.41 = 41%

3. 1.29 = 129%

Now we'll add a wrinkle. Convert the number 1.2 into a percent. Go ahead. I'll wait right here.

What did you get? Was it 120%? What you do is add a zero to 1.2 and make it 1.20, and then move the decimal two places to the right and add the percent sign. What gives you the right to add a zero? Well, it's okay to do this as long as it doesn't change the value of the number, 1.2. Since 1.2 = 1.20, you can do this. Can you add a zero to the number 30 without changing its value? Try it. Did you get 300? Does 30 = 300? If you think it does, then I'd like to trade my $30 for your $300.

Ready for another problem set? All right, then, here it comes.

PROBLEM SET

Convert each of these numbers into percents.

6. $2.6 =$ **9.** $200.1 =$

7. $1.0 =$ **10.** $45.4 =$

8. $17.3 =$

Solutions

6. $2.6 = 260\%$ **9.** $200.1 = 20,010\%$

7. $1.0 = 100\%$ **10.** $45.4 = 4,540\%$

8. $17.3 = 1,730\%$

Did you get them right? Good! Then you're ready for another wrinkle. Please convert the number 5 into a percent.

What did you get? 500%? Here's how we did it. We started with 5, added a decimal point and a couple of zeros: $5 = 5.00$. Then we converted 5.00 into a percent by moving the decimal point two places to the right and adding a percent sign: $5.00 = 500.\% = 500\%$.

Here's another group of problems for you.

PROBLEM SET

Please change each of these numbers into a percent.

11. $1 =$ **14.** $22 =$

12. $82 =$ **15.** $10 =$

13. $90 =$

Solutions

11. $1 = 100\%$ **14.** $22 = 2,200\%$

12. $82 = 8,200\%$ **15.** $10 = 1,000\%$

13. $90 = 9,000\%$

CONVERTING PERCENTS TO DECIMALS

Let's shift gears and convert some percentages into decimals.

Problem: What is the decimal equivalent of 35 percent?

Solution: $35\% = 35.0\% = .350\% = 0.35$

Problem: What is the decimal equivalent of 150 percent?

Solution: $150\% = 150.0\% = 1.500\% = 1.5$

Let's talk about what we've been doing. To convert a percent into a decimal form, drop the percent sign and move the decimal point two places to the left. In other words, do the opposite of what you did to convert a decimal into a percent.

Now I'd like you to do this problem set.

PROBLEM SET

Convert each of these percentages into decimal form.

16. $32\% =$ **19.** $603.8\% =$

17. $140\% =$ **20.** $100\% =$

18. $400\% =$

Solutions

16. $32\% = 0.32$

17. $140\% = 1.4$

18. $400\% = 4.0$ or 4

19. $603\% = 6.038$

20. $100\% = 1.0$ or 1

You may have noticed that in problem 18, I expressed the answer as 4.0 or 4. By convention, when we express a whole number, we don't use the decimal point. Similarly, in problem 20, we can drop the decimal from 1.0 and express the answer as 1.

Problem: What is the decimal equivalent of 0.3%?

Solution: $0.3\% = .003\% = .003$

Again, all you need to do is drop the percent sign and move the decimal point two places to the left. Do this problem set, so you can move on to even more exciting things.

PROBLEM SET
Convert each of these percentages into decimal form.

21. $0.95\% =$

22. $0.8\% =$

23. $0.02\% =$

24. $0.0403\% =$

25. $0.006\% =$

Solutions

21. $0.95\% = 0.0095$

22. $0.8\% = 0.008$

23. $0.02\% = 0.0002$

24. $0.0403\% = 0.000403$

25. $0.006\% = 0.00006$

NEXT STEP

So far, we've been converting decimals into percents and percents into decimals. Remember that every percent and every decimal has a fractional equivalent. So next, let's convert fractions into percents and percents into fractions.

In this lesson, you'll learn how to convert fractions into percents and percents into fractions. You'll discover some handy shortcuts and other math tricks to get you to the right answer every time.

CONVERTING FRACTIONS INTO PERCENTS AND PERCENTS INTO FRACTIONS

In the previous lesson, I said that a number could be expressed as a fraction, as a decimal, or as a percent. I said that $\frac{1}{4} = 0.25 = 25\%$. Read on to find out how this works—how fractions can be converted to percents and percents to fractions. You'll even learn more than one way to do these conversions.

CONVERTING FRACTIONS INTO PERCENTS

You may remember that in Lesson 13, you had a great time converting fractions into decimals. So $\frac{1}{4}$ is converted into 0.25 by dividing 4 into 1:

$$4 \overline{)1.0\overset{2}{0}} \quad \overset{.25}{}$$

Now let's try another way of getting from $\frac{1}{4}$ to 25%. We're going to use an old trick that I mentioned previously; it's actually a law of arithmetic. The law says that whatever you do to the bottom of a fraction, you must also do to the top. In other words, if you multiply the denominator by a certain number, you must multiply the numerator by that same number.

Let's start with the fraction $\frac{1}{4}$:

$$\frac{1 \times 25}{4 \times 25} = \frac{25}{100}$$

What did we do? We multiplied the numerator and the denominator by 25. Why 25? Because we wanted to get the denominator equal to 100. Having 100 on the bottom of a fraction makes it very easy to convert that fraction into a percent.

All right, we have $\frac{25}{100}$, which comes out to 25%. How did we do that? We removed the 100, or mathematically, we multiplied the fraction by 100, then added a percent sign. In other words,

$$\frac{25}{100} \times \frac{100}{1} = \frac{25}{{}_1 \cancel{100}} \times \frac{\cancel{100}^{1}}{1} = 25\%$$

Incidentally, *percent* means *per hundred*. Fifty-seven percent, then, means 57 per hundred. And 39 percent means 39 per hundred.

This is exactly the same process as converting a decimal into a percent. The decimal 0.25 becomes 25% when we move the decimal point two places to the right and add a percent sign. Moving a decimal two places to the right is the same as multiplying by 100. Similarly, when we changed the fraction $\frac{25}{100}$ into a percent, we also multiplied by 100 and added a percent sign.

Now you do this one.

Problem: Write $\frac{34}{100}$ as a percent.

Solution: 34%

So what you did was multiply $\frac{34}{100}$ by 100 and add a percent sign. How would you convert $\frac{9}{50}$ into a percent? Don't wait for *me* to do it. I want you to try it.

I hope you did it like this:

$$\frac{9 \times 2}{50 \times 2} = \frac{18}{100} = 18\%$$

Do you follow what I did? I multiplied the top (or numerator) by 2 and the bottom (or denominator) by 2. Am I allowed to do that? Yes! You are allowed to multiply the numerator and denominator of a fraction by the same number because it does not change that fraction's value.

Why did I multiply the numerator and denominator by 2? Because I wanted to change the denominator into 100, so that I could easily convert this fraction into a percent. So whenever you get the chance, convert the denominator into 100. It can make your life a lot easier.

PROBLEM SET
Convert these fractions into percents.

1. $\frac{6}{50} =$ **4.** $\frac{2}{25} =$

2. $\frac{7}{20} =$ **5.** $\frac{1}{5} =$

3. $\frac{8}{10} =$

Solutions

1. $\frac{6}{50} = \frac{6 \times 2}{50 \times 2} = \frac{12}{100} = 12\%$

2. $\frac{7}{20} = \frac{7 \times 5}{20 \times 5} = \frac{35}{100} = 35\%$

3. $\frac{8}{10} = \frac{8 \times 10}{10 \times 10} = \frac{80}{100} = 80\%$

4. $\frac{2}{25} = \frac{2 \times 4}{25 \times 4} = \frac{8}{100} = 8\%$

5. $\frac{1}{5} = \frac{1 \times 20}{5 \times 20} = \frac{20}{100} = 20\%$

MORE DIFFICULT CONVERSIONS

So far you've been very lucky. Every fraction has been quite easy to convert into hundredths and then, the number written over 100 is read as a percentage. For instance, $\frac{17}{100} = 17\%$ and $\frac{89}{100} = 89\%$. But what if you have a fraction that cannot easily be converted into hundredths, like $\frac{3}{8}$?

Problem: How do you change $\frac{3}{8}$ into a percent?

Solution: You do it in two steps.

First you change $\frac{3}{8}$ into a decimal:

$$8\overline{)3.\overset{6}{0}\overset{4}{0}0}\quad.375$$

Then you move the decimal point two places to the right and add a percent sign: .375 = 37.5% = 37.5%.

I did that one. Now you do this one.

Problem: Change $\frac{17}{40}$ into a percent.

Solution: 42.5%

DOING SHORT DIVISION INSTEAD OF LONG DIVISION

Do you remember the trick I showed you in Lesson 14 when we did some fast division? Dividing 40 into 17 must be done by long division, which is what I'll bet you did. However, there is a shortcut you can take. Here's my trick:

$$\frac{17}{40} = \frac{1.7}{4.0}$$

What did I do? I divided the numerator, 17, by 10, and then I divided the denominator, 40, by 10. (You can easily divide a number by 10 by simply moving its decimal point one space to the left.) But why did I bother to divide 17 and 40 by 10? Why would I rather have $\frac{1.7}{4.0}$ than $\frac{17}{40}$? Because then we can do short division instead of long division. Of course, if you happen to be using a calculator, then there is no difference between long and short division. But you're not doing that here, are you?

$$4 \overline{)1.\overset{1\,2}{7}00} \quad \overset{.425}{} = \quad 42.5\%$$

In general, when you need to divide the denominator of a fraction into the numerator, first reduce the fraction to the lowest possible terms, and then, if possible, divide the numerator and denominator by 10 or even 100 if that can get you from long division to short division.

One thing I need to mention before you do the next problem set is how to treat a repeating decimal. You'll discover that for problems 9 and 10 you'll get to the point where the same numbers keep coming up. You can divide forever and the problem never comes out even. The thing to do in this case is to stop dividing and round when you get to the tenth of a percent. When you get to the solutions for problems 9 and 10, you'll see what I mean.

Well, it's time for another problem set. Are you ready? All right, then, here it comes.

PROBLEM SET

Please change each of these fractions into percents.

6. $\frac{19}{200} =$ **9.** $\frac{37}{60} =$

7. $\frac{10}{27} =$ **10.** $\frac{13}{18} =$

8. $\frac{1}{12} =$

Solutions

6.

$$200 \overline{)19} \qquad = \quad 2 \overline{).1\overset{1}{9}0} \quad \overset{.095}{} = 9.5\%$$

7.

$$27 \overline{)10.00} \quad \overset{.37}{} = 37\%$$
$$\underline{-8\ 1X}$$
$$1\ 90$$
$$\underline{-1\ 89}$$

8.

$$12 \overline{)1.000} \quad .083 \ = 8.3\%$$
$$\underline{-96X}$$
$$40$$
$$\underline{-36}$$

9.

$$60 \overline{)37}$$

$$= 6 \overline{)3.7000} \quad .6166 \ \overset{144}{} \ = 61.7\%$$

10.

$$18 \overline{)13.000} \quad .722 \ = 72.2\%$$
$$\underline{-12\ 6XX}$$
$$40$$
$$\underline{-36}$$
$$40$$
$$\underline{-36}$$

By convention, we usually round to one decimal place. So if you rounded to a whole number or to two or three decimal places, then your answers may have differed just a bit from mine.

So how did you do? Did you get everything right? If you did, then you can pass GO, collect $200, and go directly to the next lesson. But if you didn't get all of these right, then please stay right here and work out the next set of problems. You've heard the saying "practice makes perfect." Now we'll prove it.

PROBLEM SET
Please change these fractions into percents.

11. $\frac{3}{15} =$

12. $\frac{1}{8} =$

13. $\frac{13}{22} =$

14. $\frac{19}{30} =$

15. $\frac{123}{600} =$

Solutions

11.

$$15 \overline{)3} \qquad = 5 \overline{)1.00}^{.20} \quad = 20\%$$

12.

$$8 \overline{)1.000}^{.125} \quad = 12.5\%$$

13.

$$22 \overline{)13.0000}^{.5909} \quad = 59.1\%$$

$$\begin{array}{r} -11\,0XXX \\ \hline 2\,00 \\ 1\,98 \\ \hline 200 \\ -\,198 \\ \hline 2 \end{array}$$

14.

$$30 \overline{)19} \qquad = 3 \overline{)1.9000}^{.6333} \quad = 63.3\%$$

15.

$$600 \overline{)123.000} \qquad = 6 \overline{)1.230}^{.205} \quad = 20.5\%$$

How did you make out this time? If you want still more practice, just copy each of the problems from the last two sets on another sheet of paper and work them out again.

PERCENTS INTO FRACTIONS

Now let's convert some percentages into fractions.

Problem: Convert 73% into a fraction.

Solution: $73\% = \frac{73}{100}$

Here's our instant video replay, minus the video. We dropped the percent sign and divided 73 by 100.

Problem: Convert 9% into a fraction.

Solution: $9\% = \frac{9}{100}$

Again, we dropped the percent sign and divided the 9 by 100.

Problem Set

Convert each of these percentages into fractions.

16. $46\% =$ **19.** $100\% =$

17. $10\% =$ **20.** $250\% =$

18. $7\% =$

Solutions

16. $46\% = \frac{46}{100} = \frac{23}{50}$ **19.** $100\% = \frac{100}{100} = \frac{1}{1} = 1$

17. $10\% = \frac{10}{100} = \frac{1}{10}$ **20.** $250\% = \frac{250}{100} = 2\frac{1}{2}$

18. $7\% = \frac{7}{100}$

The answers to problems 16 and 17 were reduced to their lowest possible forms. We did that too with problem 19, but by convention, we express any number divided by itself as 1. In problem 20, we reduced the improper fraction $\frac{250}{100}$ to the mixed number $2\frac{1}{2}$.

Problem: Now convert 93.6% into a fraction.

Solution: $93.6\% = \frac{93.6}{100} = \frac{936}{1000} = \frac{117}{125}$

The first step should be familiar: Get rid of the percentage sign and place 93.6 over 100. To get rid of the decimal point, we multiply the numerator, 93.6, by 10, and we multiply the denominator, 100, by 10. That gives us $\frac{936}{1000}$, which can be reduced to $\frac{117}{125}$. Sometimes we leave fractions with denominators of 100 and 1,000 as they are, even though they can be reduced. So if you leave this answer as $\frac{936}{1000}$, it's okay.

Problem: Now change 1.04% into a fraction.

Solution: $1.04\% = \frac{1.04}{100} = \frac{104}{10,000}$ or $\frac{13}{1250}$

PROBLEM SET
Change these percentages into fractions.

21. 73.5% =

24. 14.06% =

22. 1.9% =

25. 200.01% =

23. 0.8% =

Solutions

21. $73.5\% = \frac{73.5}{100} = \frac{735}{1000} = \frac{147}{200}$

22. $1.9\% = \frac{1.9}{100} = \frac{19}{1000}$

23. $0.8\% = \frac{0.8}{100} = \frac{8}{1000} = \frac{1}{125}$

24. $14.06\% = \frac{14.06}{100} = \frac{1406}{10,000} = \frac{703}{5000}$

25. $200.01\% = \frac{200.01}{100} = \frac{20,001}{10,000} = 2\frac{1}{10,000}$

For problem 25, since we don't really want to leave our answer as an improper fraction, we should convert it into a mixed number. This situation rarely comes up, so you definitely should not lose any sleep over it.

NEXT STEP

In this section, you've seen that every number has three equivalent forms—a decimal, a fraction, and a percentage. Now you can go on to finding percentage changes.

LESSON | 18

This lesson will show you how to find and understand percentage changes. You can use this knowledge to figure out many practical percentage questions that arise in your daily life.

FINDING PERCENTAGE CHANGES

I f you went to any college graduation and asked the first ten graduates you encountered to do the first problem in this lesson, chances are that no more than one or two of them would come up with the right answer. And yet percentage changes are constantly affecting us—pay increases, tax cuts, and changes in interest rates are all percentage changes. When you've completed this lesson, if someone should walk up to you and ask you to calculate a percentage change, you'll definitely be prepared.

CALCULATING PERCENTAGE CHANGE

Let's get right into it. Imagine that you were earning $500 and got a $20 raise. By what percentage did your salary go up? Try to figure it out.

We have a nice formula to help us solve percentage change problems. Here's how it works: Your salary is $500, so that's the original number. You got a $20 raise; that's the change. The formula looks like this:

$$\text{percentage change} = \frac{\text{change}}{\text{original number}}$$

Next, we substitute the numbers into the formula. And then we solve it. Once we have $\frac{20}{500}$, we could reduce it all the way down to $\frac{1}{25}$ and solve it using division:

$$= \frac{\$20}{\$500} = \frac{2}{50} = \frac{4}{100} = 4\%$$

$$\begin{array}{r} .04 \\ 25 \overline{)1.00} \\ -1.00 \end{array} = 4\%$$

Try working out this next problem on your own.

Problem: On New Year's Eve, you made a resolution to lose 30 pounds by the end of March. And sure enough, your weight dropped from 140 pounds to 110. By what percentage did your weight fall?

Solution: $\text{percentage change} = \frac{\text{change}}{\text{original number}}$

$$= \frac{30}{140} = \frac{3}{14}$$

$$\begin{array}{r} .2142 \\ 14 \overline{)3.0000} \\ -2\,8\text{XXX} \\ \hline 20 \\ -14 \\ \hline 60 \\ -56 \\ \hline 40 \\ -28 \end{array} = 21.4\%$$

PROBLEM SET

Answer the following questions using the formula shown above.

1. What is the percentage change if Becky's weight goes from 150 to 180 pounds?

2. What is the percentage change if Tom's weight goes from 130 to 200 pounds?

3. If Jessica's real estate taxes rose from $6,000 to $8,500, by what percentage did they rise?

4. Harriet's time for running a mile fell from 11 minutes to 8 minutes. By what percentage did her time fall?

Solutions

1. Percentage change $= \frac{\text{change}}{\text{original number}} = \frac{30}{150} = \frac{1}{5} = \frac{20}{100} = 20\%$

2. Percentage change $= \frac{\text{change}}{\text{original number}} = \frac{70}{130} = \frac{7}{13}$

$$
\begin{array}{r}
.538 \\
13\,\overline{)7.000} \\
-6\,5XX \\
\hline
50 \\
-39 \\
\hline
110 \\
-104 \\
\hline
6
\end{array}
\quad = 53.8
$$

3. Percentage change $= \frac{\text{change}}{\text{original number}} = \frac{\$2,500}{\$6,000} = \frac{25}{60} = \frac{5}{12}$

$$
\begin{array}{r}
.4166 \\
12\,\overline{)5.0000} \\
-4\,8XXX \\
\hline
20 \\
-12 \\
\hline
80 \\
-72 \\
\hline
80 \\
-72
\end{array}
\quad = 41.7
$$

4. Percentage change $= \frac{change}{original\ number} = \frac{3}{11}$

$$
\begin{array}{r}
.2727 \quad = 27.3\% \\
11\ )\overline{3.0000} \\
-\underline{2\ 2XXX} \\
80 \\
-\underline{77} \\
30 \\
-\underline{22} \\
80 \\
-\underline{77}
\end{array}
$$

PERCENTAGE INCREASES

Pick a number. Any number. Now triple it. By what percentage did this number increase? Take your time. Use the space in the margin to calculate the percentage.

What did you get? Three hundred percent? Nice try, but I'm afraid that's not the right answer.

I'm going to pick a number for you and then you triple it. I pick the number 100. Now I'd like you to use the percentage change formula to get the answer. (Incidentally, you may have gotten the right answer, so you may be wondering why I'm making such a big deal. But I know from sad experience that almost no one gets this right on the first try.)

So where were we? The formula. Write it down in the space below, substitute numbers into it, and then solve it.

Let's go over this problem step by step. We picked a number, 100. Next, we tripled it. Which gives us 300. Right? Now we plug some numbers into the formula. Our original number is 100. And the change when we go from 100 to 300? It's 200. From there it's just arithmetic: $\frac{200}{100} = 200\%$.

Percentage change $= \frac{change}{original\ number} = \frac{200}{100} = 200\%$

This really isn't that hard. In fact, you're going to get really good at just looking at a couple of numbers and figuring out percentage changes in your head.

Whenever you go from 100 to a higher number, the percentage increase is the difference between 100 and the new number. Suppose you were to quadruple a number. What's the percentage increase? It's 300% (400 − 100). When you double a number, what's the percentage increase? It's 100% (200 − 100).

PROBLEM SET

Here's a set of problems, and I guarantee that you'll get them all right. What's the percentage increase from 100 to each of the following?

5. 150   **8.** 500

6. 320   **9.** 425

7. 275

Solutions

5. $\frac{50}{100} = 50\%$ **8.** $\frac{400}{100} = 400\%$

6. $\frac{220}{100} = 220\%$ **9.** $\frac{325}{100} = 325\%$

7. $\frac{175}{100} = 175\%$

The number 100 is very easy to work with. Sometimes you can use it as a substitute for another number. For example, what's the percentage increase if we go from 3 to 6? Isn't it the same as if you went from 100 to 200? It's a 100% increase.

What's the percentage increase from 5 to 20? It's the same as the one from 100 to 400. It is a 300% increase.

What we've been doing here is just playing around with numbers, seeing if we can get them to work for us. As you get more comfortable with numbers, you can try to manipulate them the way we just did.

PERCENTAGE DECREASES

Remember the saying "whatever goes up must come down"? If Melanie Shapiro was earning $100 and her salary were cut to $93, by what percent was her salary cut?

Solution: The answer is obviously 7%. More formally, we divided the change in salary, $7, by the original salary, $100:

$7/$100 = 7%.

What would be the percentage decrease from 100 to 10?

Solution:
90/100 = 90%.

Here's one last problem set, and, once again, I'll guarantee that you'll get them all right.

PROBLEM SET

What is the percentage decrease from 100 to each of the following numbers?

10. 150   **11.** 20   **12.** 92   **13.** 50

Solutions

10. 39/100 = 39%   **11.** 80/100 = 80%

12. 8/100 = 8%   **13.** 50/100 = 50%

Now I'm going to throw you another curve ball. If a number—any number—were to decline by 100%, what number would you be left with? I'd really like you to think about this one.

What did you get? You should have gotten 0. That's right—no matter what number you started with, a 100% decline leaves you with 0.

NEXT STEP

Being able to calculate percentage changes is one of the most useful of all arithmetic skills. If you feel you have mastered it, then go on to the next lesson. If not, you definitely want to go back to the beginning of this lesson and make sure you get it right the second time around.

LESSON | 19

In this lesson, you'll learn how to calculate percentage distribution for several real-world scenarios. You'll find out that all percentage distributions add up to 100. You'll discover how you can check your answers after completing a problem and how to get the information you need when posed with a percentage distribution question.

PERCENTAGE DISTRIBUTION

Percentage distribution tells you the number per hundred that is represented by each group in a larger whole. For example, in Canada, 30% of the people live in cities, 45% live in suburbs, and 25% live out in the country. When you calculate percentage distributions, you'll find that they always add up to 100% (or a number very close to 100, depending on the exact decimals involved). If they don't, you'll know that you have to redo your calculations.

A class had half girls and half boys. What percentage of the class was girls, and what percentage of the class was boys? The answers are obviously 50% and 50%. That's all there is to percentage distribution. Of course the problems do get a bit more complicated, but all percentage distribution problems start out with one simple fact: The distribution will always add up to 100%.

Here's another one. One-quarter of the players on a baseball team are pitchers, one-quarter are outfielders, and the rest are infielders. What is the team's percentage distribution of pitchers, infielders, and outfielders?

Pitchers are $\frac{1}{4}$, or 25%; outfielders are also $\frac{1}{4}$, or 25%. So infielders must be the remaining 50%. Try doing the next percentage distribution on your own.

Problem: If, over the course of a week, you obtained 250 grams of protein from red meat, 150 from fish, 100 from poultry, and 50 from other sources, what percentage of your protein intake came from red meat and what percentage came from each of the other sources?

red meat	250 grams
fish	150 grams
poultry	100 grams
other	+ 50 grams
	550 grams

Try to work this out to the closest tenth of a percent. Hint: 550 grams = 100%.

Solution: red meat $= \frac{250}{550} = \frac{25}{55} = \frac{5}{11} = 45.5\%$

$$11 \overline{) 5.0000} \quad .4545$$

fish $= \frac{150}{550} = \frac{15}{55} = \frac{3}{11} = 27.3\%$

$$11 \overline{) 3.0000} \quad .2727$$

poultry $= \frac{100}{550} = \frac{10}{55} = \frac{2}{11} = 18.2\%$

$$11 \overline{) 2.0000} \quad .1818$$

other $= \frac{50}{550} = \frac{5}{55} = \frac{1}{11} = 9.1\%$

$$11 \overline{) 1.0000} \quad .0909$$

Check: $\overset{3\ 1}{45.5}$
          27.3
          18.2
        + 9.1
          100.1

When doing percentage distribution problems, it's always a good idea to check your work. If your percentages don't add up to 100 (or 99 or 101), then you've definitely made a mistake, so you'll need to go back over all your calculations. Because of rounding, my percentages added up to 100.1. Occasionally you'll end up with 100.1 or 99.9 when you check, which is fine.

Are you getting the knack? I certainly hope so because there's another problem set straight ahead.

PROBLEM SET
Calculate to the closest tenth of a percent for these problems.

1. Denver has 550,000 Caucasians, 150,000 Hispanics, 100,000 African-Americans, and 50,000 Asian-Americans. Calculate the percentage distribution of these groups living in Denver. Be sure to check your work.

2. Eleni Zimiles has 8 red beads, 4 blue beads, 3 white beads, 2 yellow beads, and 1 green bead. What is the percentage distribution of Eleni's beads?

3. Georgia-Pacific ships 5,000 freight containers a week. Fifteen hundred are sent by air, two thousand three hundred by rail, and the rest by truck. What percentage is sent by air, rail, and truck, respectively?

4. In the mayor's election Ruggerio got 45 votes, Casey got 39 votes, Schultz got 36 votes, and Jones got 28 votes. What is the percentage distribution of the vote?

5. In Middletown 65 families don't own a car; 100 families own one car; 108 families own two cars; 70 families own three cars; 40

families own four cars; and 17 families own five or more cars. What is the percentage distribution of car ownership?

Solutions

1. I got rid of the zeros (from 555,000 to 550) to make my calculation easier.

$$
\begin{array}{r}
550 \\
150 \\
100 \\
+\ 50 \\
\hline
850
\end{array}
$$

Caucasians $= \frac{550}{850} = \frac{55}{85} = \frac{11}{17}$

$$
\begin{array}{r}
.647 \quad = 64.7\% \\
17\ \overline{)\ 11.000} \\
-\ 10\ 2XX \\
\hline
80 \\
-\ 68 \\
\hline
120 \\
-\ 119 \\
\hline
\end{array}
$$

Hispanics $= \frac{150}{850} = \frac{15}{85} = \frac{3}{17}$

$$
\begin{array}{r}
.176 \quad = 17.6\% \\
17\ \overline{)\ 3.000} \\
-\ 1\ 7XX \\
\hline
130 \\
-\ 119 \\
\hline
110 \\
-\ 102 \\
\hline
8 \\
\end{array}
$$

African-Americans $= \frac{100}{850} = \frac{10}{85} = \frac{2}{17}$

$$
\begin{array}{r}
.117 \quad = 11.8\% \\
17\ \overline{)\ 2.000} \\
-\ 1\ 7XX \\
\hline
30 \\
-\ 17 \\
\hline
130 \\
-\ 119 \\
\hline
11 \\
\end{array}
$$

Asian-Americans $= \frac{50}{850} = \frac{5}{85} = \frac{1}{17}$

$$\begin{array}{r} .058 \\ 17 \overline{\smash{)}1.000} \\ -85X \\ \hline 150 \\ -136 \\ \hline 14 \end{array}$$

$= 5.9\%$

Check:

$$\begin{array}{r} \overset{2\,3}{64}.7 \\ 17.6 \\ 11.8 \\ +\ 5.9 \\ \hline 100.0 \end{array}$$

2.

$$\begin{array}{r} 8 \\ 4 \\ 3 \\ 2 \\ +1 \\ \hline 18 \end{array}$$

red $= \frac{8}{18} = \frac{4}{9}$

$$\begin{array}{r} .444 \\ 9 \overline{\smash{)}4.\overset{4\,4}{000}} \end{array}$$

$= 44.4\%$

blue $= \frac{4}{18} = \frac{2}{9}$

$$\begin{array}{r} .222 \\ 9 \overline{\smash{)}2.\overset{2\,2}{000}} \end{array}$$

$= 22.2\%$

white $= \frac{3}{18} = \frac{1}{6}$

$$\begin{array}{r} .1666 \\ 6 \overline{\smash{)}1.\overset{4\,4\,4}{0000}} \end{array}$$

$= 16.7\%$

yellow $= \frac{2}{18} = \frac{1}{9}$

$$\begin{array}{r} .111 \\ 9 \overline{\smash{)}1.\overset{1\,1}{000}} \end{array}$$

$= 11.1\%$

green $= \frac{1}{18}$

$$\begin{array}{r} .0555 \\ 18 \overline{\smash{)}1.0000} \\ -90XX \\ \hline 100 \\ -90 \\ \hline 100 \\ -90 \\ \hline 10 \end{array}$$

$= 5.6\%$

Check: $\overset{2\,2}{44.4}$

22.2

16.7

11.1

$+\ 5.6$

100.0

3. air $= \frac{1500}{5000} = \frac{15}{50} = \frac{30}{100} = 30\%$

rail $= \frac{2300}{5000} = \frac{23}{50} = \frac{46}{100} = 46\%$

truck $= \frac{1200}{5000} = \frac{12}{50} = \frac{24}{100} = 24\%$

Check: $\overset{1}{30}$

46

$+\ 24$

100

4. $\overset{2}{45}$

39

36

$+\ 28$

148

Ruggerio $= \frac{45}{148}$

$$\begin{array}{r} .304 \quad = 30.4\% \\ 148\ \overline{)45.000} \\ -\ 44\ 4\text{XX} \\ \hline 600 \\ -\ 592 \\ \hline \end{array}$$

Casey $= \frac{39}{148}$

$$\begin{array}{r} .263 \quad = 26.3\% \\ 148\ \overline{)39.000} \\ -\ 29\ 6\text{XX} \\ \hline 9\ 40 \\ -\ 8\ 88 \\ \hline 520 \\ -\ 444 \\ \hline 76 \end{array}$$

Schultz $\quad = \frac{36}{148} = \frac{9}{37}$

$$\begin{array}{r} .243 \quad = 24.3\% \\ 37 \overline{)9.000} \\ \underline{- 7\ 4XX} \\ 1\ 60 \\ \underline{- 1\ 48} \\ 120 \\ \underline{- 111} \\ 9 \end{array}$$

Jones $\quad = \frac{28}{148} = \frac{7}{37}$

$$\begin{array}{r} .189 \quad = 18.9\% \\ 37 \overline{)7.000} \\ \underline{- 3\ 7XX} \\ 330 \\ \underline{- 296} \\ 340 \\ \underline{- 333} \\ 7 \end{array}$$

Check:

$$\begin{array}{r} ^{1\ 1}30.4 \\ 26.3 \\ 24.3 \\ + \underline{18.9} \\ 99.9 \end{array}$$

5.

$$\begin{array}{r} ^{2}65 \\ ^{2}100 \\ 108 \\ 70 \\ 40 \\ + \underline{17} \\ 400 \end{array}$$

no cars $\quad = \frac{65}{400} = \frac{13}{80}$

$$\begin{array}{r} .162 \quad = 16.3\% \\ 80 \overline{)13.000} \\ \underline{- 8\ 0XX} \\ 5\ 000 \\ \underline{- 4\ 800} \\ 200 \\ \underline{- 160} \\ 40 \end{array}$$

177

1 car $= \frac{100}{400} = \frac{1}{4} = .25 = 25\%$

2 cars $= \frac{108}{400} = \frac{27}{100} = 27\%$

3 cars $= \frac{70}{400} = \frac{7}{40}$

$$
\begin{array}{r}
.175 \quad = 17.5\% \\
40 \overline{)7.000} \\
-4\,0\text{XX} \\
\hline
3\,00 \\
-2\,80 \\
\hline
200 \\
-200 \\
\hline
\end{array}
$$

4 cars $= \frac{40}{400} = \frac{10}{100} = 10.0\%$

5 or more cars $= \frac{17}{400}$

$$
\begin{array}{r}
.042 \quad = 4.3\% \\
400 \overline{)17.000} \\
-16\,00\text{X} \\
\hline
1\,000 \\
-800 \\
\hline
200 \\
\end{array}
$$

Check:

$$
\begin{array}{r}
{}^{3\,1}16.3 \\
25.0 \\
27.0 \\
17.5 \\
10.0 \\
+\,4.3 \\
\hline
100.1 \\
\end{array}
$$

NEXT STEP

Congratulations on learning how to calculate percentage distribution. The next lesson shows how to find percentages of numbers. Go for it!

LESSON | 20

In this lesson, you'll learn how to find percentages of numbers. We'll start with the percentage of your pay that the Internal Revenue Service collects.

FINDING PERCENTAGES OF NUMBERS

The Internal Revenue Service charges different tax rates for different levels of income. For example, most middle-income families are taxed at a rate of 28 percent on some of their income. Suppose that one family had to pay 28 percent of $10,000. How much would that family pay?

Solution:

$10,000 \times .28 = \$2,800$

You'll notice that we converted 28 percent into the decimal .28 to carry out that calculation. We used fast multiplication, which we covered in Lesson 14.

Problem: How much is 14.5 percent of 1,304?

Solution:
$$
\begin{array}{r}
1{,}304 \\
\times\ .145 \\
\hline
6520 \\
5216 \\
1304 \\
\hline
189.080
\end{array}
$$

Problem: How much is 73.5 percent of $12,416.58?

Solution:
$$
\begin{array}{r}
\$12{,}416.58 \\
\times\ .735 \\
\hline
6208290 \\
3724974 \\
8691606 \\
\hline
\$9{,}126.18630\ =\ \$9{,}126.19
\end{array}
$$

Most people try to leave around a 15 percent tip in restaurants. In New York City, where the sales tax is 8.25 percent, customers often just double the tax. But there is actually another very fast and easy way to calculate that 15 percent tip.

Let's say that your bill comes to $28.19. Round it off to $28, the nearest even dollar amount. Then find 10 percent of $28, which is $2.80. Now what's half of $2.80? It's $1.40. How much is $1.40 plus $2.80? It's $4.20.

Let's try a much bigger check—$131.29. Round it off to the nearest even dollar amount—$132. What is 10 percent of $132? It's $13.20. And how much is half of $13.20? It's $6.60. Finally, add $13.20 and $6.60 together to get your $19.80 tip.

PROBLEM SET

1. How much is 13 percent of 150?

2. Find 34.5 percent of $100.

3. How much is 22.5 percent of $390?

4. Find 78.2 percent of $1,745.

5. Find 56.3 percent of 1,240.

6. How much is 33.8 percent of $29,605.28?

Solutions

1.

```
      150
    × .13
      450
      150
    19.5̶0̶
```

4.

```
    $1,745
    × .782
      3490
     13960
     12215
  $1,364.59̶0̶
```

2. $100 × .345 = $34.50

5.

```
    1,240
    × .563
     3720
     7440
     6200
   698.12̶0̶
```

3.

```
      $390
    × .225
      1950
       780
       780
   $87.75̶0̶
```

6.

```
    $29,605.28
       × .338
     23684224
      8881584
      8881584
  $10,006.58̶4̶6̶4̶
```

NEXT STEP

You've already done some applications, so the next lesson will be easy. Let's find out.

LESSON | 21

In this lesson, you'll be able to pull together all of the math skills you've mastered in this section and apply them to situations you may encounter in your daily life. You'll see how practical the knowledge that you've gained is and how often mathematical questions arise that you now know the answers to.

APPLICATIONS

Can you believe it? You're about to begin the last lesson in this book. Many of the problems you'll be solving here are ones you may encounter at work, at home, or while driving or shopping.

Before you get started, how about a few practice problems? First, a markup problem. Stores pay one price for an item, but they almost always charge a higher price to their customers. We call that process a markup. For instance, if a store owner pays $10 for a radio and sells it for $15, by what percentage did she mark it up?

Percentage markup $= \frac{\$5}{\$10} = .50 = 50\%$

Markdown is another common commercial term. Suppose a store advertises that every item is marked down by 40%. If a CD was originally selling for $8, what would its marked-down price be during the sale?

$$\text{Sale price} = \$8 - (\$8 \times 0.40)$$
$$= \$8 - \$3.20$$
$$= \$4.80$$

Try this next one yourself.

Problem: Imagine that you're earning $250 a week and receive a raise of 10 percent. How much is your new salary?

Solution: New salary $= \$250 + (\$250 \times 0.10)$
$= \$250 + \$25$
$= \$275$

Problem: Suppose you went on a big diet, and your weight fell by 20%. If you started out weighing 150 pounds, how much would you weigh after dieting?

Solution: New weight $= 150 - (150 \times 0.20)$
$= 150 - 30$
$= 120$

Here's another type of problem.

Problem: What percentage of 100 is 335?

Solution: $\frac{335}{100} = 3.35 = 335\%$

Congratulations! You've just gotten a $100 salary increase. How much of that $100 do you actually take home if you have to pay 15 percent in federal income tax, 7.65 percent in payroll tax, and 2.5 percent in state income tax?

Solution:

$$15.00\%$$
$$7.65$$
$$+ \ 2.50$$
$$25.25\%$$

$100 \times 25.25\% = \$25.25$ taxes paid

$$\$100.00$$
$$- \ 25.25$$
$$\$74.75 \ = \text{money you take home}$$

PROBLEM SET

Let's apply what you've learned about percentages to some real-life situations. Check your work with the solutions at the end of the lesson.

1. The Happy Day Nursing Home increased the number of beds from 47 to 56. By what percentage did they increase?

2. If one-quarter of all Americans live in cities, what is the percentage of Americans who do not live in cities?

3. Three people ran for state Senator. If Marks got one-third of the vote and Brown got one-fifth of the vote, what percentage of the vote did Swanson receive?

4. If you had four pennies, two nickels, three dimes, and a quarter, what percentage of a dollar would you have?

5. If 8.82 is the average score in a swim meet and you had a score of 9.07, by what percentage did your score exceed the average?

6. A dress is marked up 65% from what it cost the store owner. If the store owner paid $20 for the dress, how much does she charge?

7. A suit on sale is marked down 40% from its regular price. If its regular price is $170, how much is its sale price?

8. Of 319 employees at the Smithtown Mall, 46 were out sick. What percentage of employees were at work that day?

9. Henry Jones gets a hit 32.8% of the times he comes to bat. What is his batting average? (Hint: a batting average is a decimal expressed in thousandths.)

10. You're driving at 40 mph and increase your speed by 20%. How fast are you now going?

11. You cut back on eating and your $50 weekly food bill falls by 30%. What is your new food bill?

12. Jason Jones was getting 20 miles per gallon. But when he slowed down to an average speed of 70 mph, his gas mileage rose by 40%. What is his new gas mileage?

13. If you were making $20,000 and got a 15% pay increase, how much would you now be making?

14. A school that had 650 students had a 22% increase in enrollment. How much is its new enrollment?

15. What would your percentage score on an exam be if you got 14 questions right out of a total of 19 questions?

16. What percentage of a dollar is $4.58?

17. If you needed $500 and had saved $175, what percentage of the $500 had you saved?

18. The University of Wisconsin alumni association has 45,000 members. Four thousand five hundred are women 40 and under; seven thousand nine hundred are women over 40; twelve thousand eight hundred are men 40 and under; the remainder of members consists of men over 40. Find the percentage distribution of all four membership categories. Remember to check your work.

19. Mr. Philips baked three apple pies, two blueberry pies, five cherry pies, and six key lime pies for the town bake-off. What percentage of the pies were apple, blueberry, cherry, and key lime?

20. During the July 4th weekend, a video store rented out 300 westerns, 450 martial arts movies, 100 musicals, 250 children's movies, and 50 foreign films. What percentage of the rentals was in each category?

Solutions

1. $\frac{9}{47} =$

$$
\begin{array}{r}
.191 \quad = 19.1\% \\
\overline{)\,9.000} \\
-4\ 7XX \\
\hline
4\ 30 \\
-4\ 23 \\
\hline
70 \\
-47 \\
\hline
23
\end{array}
$$

2. $\frac{3}{4} = \frac{75}{100} = 75\%$

3. $1 - (\frac{1}{3} + \frac{1}{5}) = 1 - (\frac{1 \times 5}{3 \times 5} + \frac{1 \times 3}{5 \times 3}) = 1 - (\frac{5}{15} + \frac{3}{15}) = 1 - \frac{8}{15} = \frac{7}{15}$

$$
\begin{array}{r}
.466 \quad = 46.7\% \\
15\ \overline{)\,7.000} \\
-6\ 0XX \\
\hline
100 \\
-90 \\
\hline
100 \\
-90 \\
\hline
10
\end{array}
$$

4.

$$
\begin{array}{r}
\$0.04 \\
0.10 \\
0.30 \\
+\ 0.25 \\
\hline
\$0.69
\end{array}
$$
$\frac{69}{100} = 69\%$

5.

$$9.07$$
$$\underline{-8.82}$$
$$0.25$$

$$\frac{0.25}{8.82} = \frac{25}{882} =$$

$$\begin{array}{r} .0283 \\ 882 \overline{)25.0000} \\ \underline{17\ 64\text{xx}} \\ 7\ 360 \\ \underline{-7\ 056} \\ 3040 \\ \underline{-2646} \end{array} \quad = 2.83\%$$

6. price

= $20 + ($20 × 0.65)
= $20 + $13
= $33

$$\begin{array}{r} 0.65 \\ \underline{\times\ \$20} \\ \$13.00 \end{array}$$

7. sale price

= $170 − ($170 × 0.4)
= $170 − $68
= $102

$$\begin{array}{r} \$170 \\ \underline{\times\ 0.4} \\ \$68.0 \end{array}$$

8.

$$319$$
$$\underline{-46}$$
$$273$$

$$\frac{273}{319} =$$

$$\begin{array}{r} .855 \\ 319 \overline{)273.000} \\ \underline{-255\ 2\text{xx}} \\ 17\ 80 \\ \underline{-15\ 95} \\ 1\ 850 \\ \underline{-1\ 595} \\ 255 \end{array} \quad = 85.6\%$$

9. 32.8% = 0.328

10. We converted 20% to 0.2 in order to work out this problem.

40 + (0.2 × 40) = 40 + 8 = 48 mph

11. $50 − (0.3 × $50) = $50 − $15 = $35

12. 20 + (0.4 × 20) = 20 + 8 = 28 miles per gallon

13. $20,000 + (0.15 × $20,000) = $20,000 + $3,000 = $23,000

14. $650 + (0.22 \times 650) = 650 + 143 = 793$

$$
\begin{array}{r}
650 \\
\times\,0.22 \\
\hline
13\,00 \\
130\,0 \\
\hline
143.00
\end{array}
$$

15. $\frac{14}{19} =$

$$
\begin{array}{r}
.736 \\
19\,\overline{)\,14.000} \\
-13\,3XX \\
\hline
70 \\
-57 \\
\hline
130 \\
-114 \\
\hline
16
\end{array}
$$

$= 73.7\%$

16. $\frac{\$4.58}{\$1.00} = 4.58 = 458\%$

17. $\frac{\$175}{\$500} = \frac{\$350}{\$1,000} = 35\%$

18. women 40 and under $= \frac{4,500}{45,000} = \frac{45}{450} = \frac{1}{10} = \frac{10}{100} = 10\%$

women over 40 $= \frac{7,900}{45,000} = \frac{79}{450} = \frac{7.9}{45} =$

$$
\begin{array}{r}
.175 \\
45\,\overline{)\,7.900} \\
4\,5XX \\
\hline
3\,40 \\
-3\,15 \\
\hline
250 \\
-225 \\
\hline
25
\end{array}
$$

$= 17.6\%$

men 40 and under $= \frac{12,800}{45,000} = \frac{128}{450} = \frac{64}{225} =$

$$
\begin{array}{r}
.284 \\
225\,\overline{)\,64.000} \\
-45\,XXX \\
\hline
19\,00 \\
-18\,00 \\
\hline
1\,000 \\
-900 \\
\hline
100
\end{array}
$$

$= 28.4\%$

$$4,500$$
$$7,900$$
$$\underline{+\ 12,800}$$
$$25,200$$

$$45,000$$
$$\underline{-\ 25,200}$$
$$19,800$$

men over 40 $= \frac{19,800}{45,000} = \frac{198}{450} = \frac{99}{225} = \frac{33}{75} = \frac{11}{25}$

$$\begin{array}{r} .44 \\ 25\overline{)11.00} \\ \underline{-\ 10\ 0\text{X}} \\ 1\ 00 \\ \underline{-\ 1\ 00} \end{array} \quad = 44\%$$

Check:
$$10.0$$
$$17.6$$
$$28.4$$
$$\underline{+\ 44.0}$$
$$100.0$$

19. apple $= \frac{3}{16} =$

$$\begin{array}{r} .1875 \\ 16\overline{)3.0000} \\ \underline{-\ 1\ 6\text{XXX}} \\ 1\ 40 \\ \underline{-\ 1\ 28} \\ 120 \\ \underline{-\ 112} \\ 80 \\ \underline{-\ 80} \end{array} \quad = 18.8\%$$

blueberry $= \frac{2}{16} = \frac{1}{8}$

$$\begin{array}{r} .125 \\ 8\overline{)1.0\overset{2\ 4}{0}0} \end{array} \quad = 12.5\%$$

cherry $= \frac{5}{16} =$

$$\begin{array}{r} .3125 \quad = 31.3\% \\ 16\overline{)5.0000} \\ \underline{-48XXX} \\ 20 \\ \underline{-16} \\ 40 \\ \underline{-32} \\ 80 \\ \underline{-80} \end{array}$$

key lime $= \frac{6}{16} = \frac{3}{8}$

$$\begin{array}{r} .375 \quad = 37.5\% \\ 8\overline{)3.0\overset{6}{0}\overset{4}{0}} \end{array}$$

Check:
$$\begin{array}{r} 18.8 \\ 12.5 \\ 31.3 \\ \underline{+37.5} \\ 100.1 \end{array}$$

20.
$$\begin{array}{r} 300 \\ 450 \\ 100 \\ 250 \\ \underline{+50} \\ 1,150 \end{array}$$

westerns $= \frac{300}{1,150} = \frac{30}{115} = \frac{6}{23} =$

$$\begin{array}{r} .260 \quad = 26.1\% \\ 23\overline{)6.0000} \\ \underline{-46XXX} \\ 1\,40 \\ \underline{-1\,38} \\ 200 \end{array}$$

martial arts $= \frac{450}{1,150} = \frac{45}{115} = \frac{9}{23} =$

$$
\begin{array}{r}
.391 \\
23\overline{\smash{\big)}9.0000} \\
\underline{-6\ 9XXX} \\
2\ 10 \\
\underline{-2\ 07} \\
30 \\
\underline{-23} \\
70
\end{array}
$$

$= 39.1\%$

musicals $= \frac{100}{1,150} = \frac{10}{115} = \frac{2}{23} =$

$$
\begin{array}{r}
.086 \\
23\overline{\smash{\big)}2.0000} \\
\underline{-1\ 84XX} \\
160 \\
\underline{-138} \\
220
\end{array}
$$

$= 8.7\%$

children's movies $= \frac{250}{1,150} = \frac{25}{115} = \frac{5}{23} =$

$$
\begin{array}{r}
.217 \\
23\overline{\smash{\big)}5.000} \\
\underline{-4\ 6XX} \\
40 \\
\underline{-23} \\
170 \\
\underline{-161} \\
9
\end{array}
$$

$= 21.7\%$

foreign films $= \frac{50}{1,150} = \frac{5}{115} = \frac{1}{23} =$

$$
\begin{array}{r}
.043 \\
23\overline{\smash{\big)}1.0000} \\
\underline{-92XX} \\
80 \\
\underline{-69} \\
110
\end{array}
$$

$= 4.3$

Check:

$$
\begin{array}{r}
26.1 \\
39.1 \\
8.7 \\
21.7 \\
\underline{+\ 4.3} \\
99.9
\end{array}
$$

If a storewide sale sounds too good to be true, it probably is. Like this one: "All prices reduced by 50%. On all clothing, take off an additional 30%. And on item with red tags, take off an additional 25%."

OK, doesn't that mean that on red-tagged clothing you take off 105%, which means that on a dress originally priced at $100, the store gives you $5 to take it off their hands? Evidently not.

How much would you actually have to pay for that dress?

Solution

First take off 50%: $100 × .50 = $50.

Now take off another 30%: $50 × .70 = $35. Notice the shortcut we just took. Instead of multiplying $50 × .30, getting $15, and subtracting $15 from $50 to get $35, we saved ourselves a step by multiplying $50 by .70.

Finally, we take off another 25%: $35 × .75 = $26.25. Notice that we take the same shortcut, instead of multiplying $35 by .25, and subtracting $8.75 from $35.

To summarize, we take 50% off the original $100 price, then 30% off the new $50 price, and then 25% off the price of $35. A price reduction from $100 to $26.25 is not too shabby, but that's a far cry from a reduction of $105.

NEXT STEP

Congratulations on completing the 21 lessons in this book! You deserve a break. Don't look now, but there is one more chance for you to exercise the skills you've learned thus far. After you've taken a well-deserved break, check out the final exam.

FINAL EXAM

You didn't think you'd actually be able to get out of here without taking a final exam, did you? If you know this stuff cold, then this exam will be a piece of cake. And if you don't do so well, you'll see exactly where you need work, and you can go back to those specific lessons so that you can master those concepts.

REVIEW LESSON 1

Add these columns of numbers.

1.
596
372
952
183
465
+ 238

3.
12,695
10,483
15,752
11,849
17,304
+ 20,176

2.
1,906
2,734
1,075
3,831
+ 4,570

REVIEW LESSON 2

Subtract these numbers.

4.
463
− 165

6.
11,401
− 9,637

5.
1,432
− 1,353

REVIEW LESSON 3

Multiply these numbers.

7.
339
× 276

9.
15,773
× 16,945

8.
4,715
× 3,896

REVIEW LESSON 4

Perform each of these divisions.

10. $8\overline{)14{,}173}$

12. $536\overline{)1{,}734{,}613}$

11. $29\overline{)310{,}722}$

LESSON 1

Convert these improper fractions into mixed numbers.

13. $\frac{22}{6} =$

15. $\frac{79}{4} =$

14. $\frac{86}{7} =$

Convert these mixed numbers into improper fractions.

16. $2\frac{5}{8} =$

18. $7\frac{5}{9} =$

17. $9\frac{1}{4} =$

LESSON 2

Add these fractions.

19. $\frac{1}{2} + \frac{1}{6} + \frac{1}{4} =$

20. $\frac{2}{3} + \frac{5}{8} + \frac{3}{4} =$

21. $\frac{2}{7} + \frac{4}{5} + \frac{5}{6} =$

LESSON 3

Subtract each of these fractions.

22. $\frac{7}{8} - \frac{1}{4} =$

23. $\frac{4}{5} - \frac{2}{7} =$

24. $\frac{8}{9} - \frac{3}{5} =$

LESSON 4

Perform these multiplications.

25. $\frac{3}{7} \times \frac{4}{5} =$

26. $\frac{3}{4} \times \frac{7}{8} =$

27. $\frac{2}{3} \times \frac{5}{6} =$

LESSON 5

Perform these divisions.

28. $\frac{1}{8} \div \frac{1}{4} =$

29. $\frac{5}{7} \div \frac{5}{9} =$

30. $\frac{2}{3} \div \frac{5}{8} =$

LESSON 6

Perform the operations indicated with these fractions.

31. $\frac{8}{3} + \frac{7}{4} =$

32. $\frac{5}{2} + \frac{7}{6} =$

33. $\frac{9}{4} - \frac{3}{2} =$

34. $\frac{17}{5} - \frac{4}{3} =$

35. $\frac{10}{3} \times \frac{5}{4} =$

36. $\frac{23}{5} \times \frac{10}{7} =$

37. $\frac{25}{4} \div \frac{10}{8} =$

38. $\frac{37}{5} \div \frac{12}{7} =$

Lesson 7

Perform the operations indicated with these mixed numbers.

39. $2\frac{2}{3} + 1\frac{3}{4} =$

40. $5\frac{1}{8} + 3\frac{2}{5} =$

41. $3\frac{7}{8} - 2\frac{1}{4} =$

42. $6\frac{2}{3} - 4\frac{1}{2} =$

43. $2\frac{1}{4} \times 3\frac{2}{7} =$

44. $1\frac{5}{6} \times 3\frac{1}{3} =$

45. $6\frac{5}{7} \div 2\frac{7}{9} =$

46. $5\frac{1}{4} \div 2\frac{2}{3} =$

Lesson 8

Do each of these problems.

47. Three babies were born on the same day. The first weighed $7\frac{1}{2}$ pounds, the second weighed $6\frac{3}{4}$ pounds, and the third weighed $7\frac{7}{8}$ pounds. How much did the three babies weigh all together?

48. Ellen did a running broad jump of 16 feet, $9\frac{1}{4}$ inches. Joan did a jump of 16 feet $5\frac{3}{8}$ inches. How much farther did Ellen jump?

49. A large cake was shared equally by three families. If each family had four members, what fraction of the cake did each person receive?

LESSON 9
Perform these additions and subtractions.

50. $1.04 + 3.987 =$

53. $100.66 + 299.54 =$

51. $7.909 + 16.799 =$

54. $46.3 - 19.42 =$

52. $15.349 + 6.87 =$

55. $104.19 - 55.364 =$

LESSON 10
Do each of these multiplication problems.

56.
$$\begin{array}{r} 3.96 \\ \times\, 1.53 \\ \hline \end{array}$$

58.
$$\begin{array}{r} 10.70 \\ \times\, 19.52 \\ \hline \end{array}$$

57.
$$\begin{array}{r} 18.56 \\ \times\, 13.08 \\ \hline \end{array}$$

LESSON 11
Do each of these division problems.

59. $\frac{1.06}{0.87} =$

61. $\frac{90}{4.17} =$

60. $\frac{12}{1.16} =$

LESSON 12
Express each of these numbers as a fraction and as a decimal.

62. four hundredths

63. thirty-one thousandths

64. six hundred ninety-one thousandths

LESSON 13

Convert these fractions into decimals.

65. $\frac{4}{5} =$

67. $\frac{48}{200} =$

66. $\frac{9}{15} =$

Convert these decimals into fractions.

68. $0.93 =$

70. $0.47 =$

69. $0.003 =$

LESSON 14

Multiply each of these numbers by 1,000.

71. 0.03

73. 0.092

72. 1.5

Divide each of these numbers by 100.

74. 6

76. 0.004

75. 0.1

LESSON 15

Work out each of these problems.

77. If you had six quarters, nine dimes, fifteen nickels, and eight pennies, how much money would you have?

78. Carpeting costs $8.99 a yard. If Mark buys 16.2 yards, how much will this cost him?

79. If you bought $4\frac{1}{4}$ pounds of peanuts at $1.79 a pound and $4\frac{1}{2}$ pounds of cashews at $4.50 a pound, how much would you spend all together?

LESSON 16

Convert these decimals into percents.

80. $0.9 =$ **82.** $0.07 =$

81. $1.62 =$

Convert these percents into decimals.

83. $20\% =$ **85.** $4\% =$

84. $150\% =$

LESSON 17

Convert these fractions into percents.

86. $\frac{7}{10} =$ **88.** $\frac{91}{100} =$

87. $\frac{3}{8} =$

Convert these percents into fractions.

89. $14\% =$ **91.** $13.9\% =$

90. $200\% =$

LESSON 18

Work out each of these problems.

92. You were earning $400 a week and got a $50 raise. By what percent did your salary increase?

93. Your weight fell from 180 pounds to 160 pounds. By what percent did your weight decrease?

94. What is the percentage change if we go from 225 to 250?

LESSON 19

Work out each of these problems.

95. John has four blue marbles, three red marbles, two green marbles, and 1 yellow marble. What is his percentage distribution of red, blue, green, and yellow marbles?

96. If Marsha received one-third of the vote, Bill received two-fifths of the vote, and Diane received the rest of the votes in a class election, what percent of the votes did Diane receive?

LESSONS 20 AND 21

Work out each of these problems.

97. If you had six pennies, three nickels, four dimes, and a quarter, what percentage of a dollar would you have?

98. You're driving at 50 mph and increase your speed by 20%. How fast are you now going?

99. If your restaurant bill came to $16.85 and you left a 15% tip, how much money would you leave for the tip?

Solutions

1.
```
  4 2
  596
  372
  952
  183
  465
+ 238
2,806
```

2.
```
  3 2 1
  1,906
  2,734
  1,075
  3,831
+ 4,570
 14,116
```

3.
```
  1 3 3 2
  12,695
  10,483
  15,752
  11,849
  17,304
+ 20,176
  88,259
```

4.
```
  3 15 1
  463
- 165
  298
```

5.

$$1,\overset{3}{\cancel{4}}\overset{12}{\cancel{3}}\overset{1}{2}$$
$$-\ 1,353$$
$$79$$

6.

$$\overset{10}{\cancel{1}}\overset{13}{\cancel{1}},\overset{9}{\cancel{4}}\overset{1}{\cancel{0}}1$$
$$-\ 9,637$$
$$1,764$$

7.

$$339$$
$$\times\ 276$$
$$2\ 034$$
$$23\ 73$$
$$67\ 8\ \ $$
$$93,564$$

8.

$$4,715$$
$$\times\ 3,896$$
$$28\ 290$$
$$424\ 35$$
$$3\ 772\ 0$$
$$14\ 145\ \ \ $$
$$18,369,640$$

9.

$$15,773$$
$$\times\ 16,945$$
$$78\ 865$$
$$630\ 92$$
$$14\ 195\ 7$$
$$94\ 638$$
$$157\ 73\ \ \ $$
$$267,273,485$$

10.

$$8\ )\overline{14,\overset{6\ 5\ 1}{173}}\quad 1,771\ \ R5$$

11.

$$29\ )\overline{310,722}\quad 10,714\ \ R16$$
$$-\ 29X\ XXX$$
$$20\ 7$$
$$-\ 20\ 3\ \ $$
$$42$$
$$-\ 29\ \ $$
$$132$$
$$-\ 116$$
$$16$$

12.

$$536\ )\overline{1,734,613}\quad 3,236\ \ R117$$
$$-\ 1,608\ XXX$$
$$126\ 6$$
$$-\ 107\ 2\ \ $$
$$19\ 41$$
$$-\ 16\ 08\ \ $$
$$3\ 333$$
$$-\ 3\ 216$$
$$117$$

13. $\frac{22}{6}=3\frac{4}{6}=3\frac{2}{3}$

14. $\frac{86}{7}=12\frac{2}{7}$

15. $\frac{79}{4}=19\frac{3}{4}$

16. $2\frac{5}{8}=\frac{21}{8}$

17. $9\frac{1}{4}=\frac{37}{4}$

18. $7\frac{5}{9}=\frac{68}{9}$

19. $\frac{1}{2} + \frac{1}{6} + \frac{1}{4} = \frac{1 \times 6}{2 \times 6} + \frac{1 \times 2}{6 \times 2} + \frac{1 \times 3}{4 \times 3} = \frac{6}{12} + \frac{2}{12} + \frac{3}{12} = \frac{11}{12}$

20. $\frac{2}{3} + \frac{5}{8} + \frac{3}{4} = \frac{2 \times 8}{3 \times 8} + \frac{5 \times 3}{8 \times 3} + \frac{3 \times 6}{4 \times 6} = \frac{16}{24} + \frac{15}{24} + \frac{18}{24} = \frac{49}{24} = 2\frac{1}{24}$

21. $\frac{2}{7} + \frac{4}{5} + \frac{5}{6} = \frac{2 \times 30}{7 \times 30} + \frac{4 \times 42}{5 \times 42} + \frac{5 \times 35}{6 \times 35} = \frac{60}{210} + \frac{168}{210} + \frac{175}{210} = \frac{403}{210} = 1\frac{193}{210}$

22. $\frac{7}{8} - \frac{1}{4} = \frac{7}{8} - \frac{1 \times 2}{4 \times 2} = \frac{7}{8} - \frac{2}{8} = \frac{5}{8}$

23. $\frac{4}{5} - \frac{2}{7} = \frac{4 \times 7}{5 \times 7} - \frac{2 \times 5}{7 \times 5} = \frac{28}{35} - \frac{10}{35} = \frac{18}{35}$

24. $\frac{8}{9} - \frac{3}{5} = \frac{8 \times 5}{9 \times 5} + \frac{3 \times 9}{5 \times 9} = \frac{40}{45} - \frac{27}{45} = \frac{13}{45}$

25. $\frac{3}{7} \times \frac{4}{5} = \frac{12}{35}$

26. $\frac{3}{4} \times \frac{7}{8} = \frac{21}{32}$

27. $\frac{2}{3}^{1} \times \frac{5}{3\!6} = \frac{5}{9}$

28. $\frac{1}{8} \div \frac{1}{4} = \frac{1}{2\!8} \times \frac{4}{1}^{1} = \frac{1}{2}$

29. $\frac{5}{7} \div \frac{5}{9} = \frac{5}{7}^{1} \times \frac{9}{1\!5} = \frac{9}{7} = 1\frac{2}{7}$

30. $\frac{2}{3} \div \frac{5}{8} = \frac{2}{3} \times \frac{8}{5} = \frac{16}{15} = 1\frac{1}{15}$

31. $\frac{8}{3} + \frac{7}{4} = \frac{8 \times 4}{3 \times 4} + \frac{7 \times 3}{4 \times 3} = \frac{32}{12} + \frac{21}{12} = \frac{53}{12} = 4\frac{5}{12}$

32. $\frac{5}{2} + \frac{7}{6} = \frac{5 \times 3}{2 \times 3} + \frac{7}{6} = \frac{15}{6} + \frac{7}{6} = \frac{22}{6} = 3\frac{4}{6} = 3\frac{2}{3}$

33. $\frac{9}{4} - \frac{3}{2} = \frac{9}{4} - \frac{3 \times 2}{2 \times 2} = \frac{9}{4} - \frac{6}{4} = \frac{3}{4}$

34. $\frac{17}{5} - \frac{4}{3} = \frac{17 \times 3}{5 \times 3} - \frac{4 \times 5}{3 \times 5} = \frac{51}{15} - \frac{20}{15} = \frac{31}{15} = 2\frac{1}{15}$

35. $\frac{10}{3}^{5} \times \frac{5}{2\!4} = \frac{25}{6} = 4\frac{1}{6}$

36. $\frac{23}{1\!5} \times \frac{10}{7}^{2} = \frac{46}{7} = 6\frac{4}{7}$

37. $\frac{25}{4} \div \frac{10}{8} = \frac{25}{1\!4}^{5} \times \frac{8}{2\!10}^{2} = \frac{10}{2} = 5$

38. $\frac{37}{5} \div \frac{12}{7} = \frac{37}{5} \times \frac{7}{12} = \frac{259}{60} = 4\frac{19}{60}$

39. $2\frac{2}{3} + 1\frac{3}{4} = \frac{8}{3} + \frac{7}{4} = \frac{8 \times 4}{3 \times 4} + \frac{7 \times 3}{4 \times 3} = \frac{32}{12} + \frac{21}{12} = \frac{53}{12} = 4\frac{5}{12}$

40. $5\frac{1}{8} + 3\frac{2}{5} = \frac{41}{8} + \frac{17}{5} = \frac{41 \times 5}{8 \times 5} + \frac{17 \times 8}{5 \times 8} = \frac{205}{40} + \frac{136}{40} = \frac{341}{40} = 8\frac{21}{40}$

41. $3\frac{7}{8} - 2\frac{1}{4} = \frac{31}{8} - \frac{9}{4} = \frac{31}{8} - \frac{9 \times 2}{4 \times 2} = \frac{31}{8} - \frac{18}{8} = \frac{13}{8} = 1\frac{5}{8}$

42. $6\frac{2}{3} - 4\frac{1}{2} = \frac{20}{3} - \frac{9}{2} = \frac{20 \times 2}{3 \times 2} - \frac{9 \times 3}{2 \times 3} = \frac{40}{6} - \frac{27}{6} = \frac{13}{6} = 2\frac{1}{6}$

43. $2\frac{1}{4} \times 3\frac{2}{7} = \frac{9}{4} \times \frac{23}{7} = \frac{207}{28} = 7\frac{11}{28}$

44. $1\frac{5}{6} \times 3\frac{1}{3} = \frac{11}{\cancel{6}\,_3} \times \frac{\cancel{10}^{\,5}}{3} = \frac{55}{9} = 6\frac{1}{9}$

45. $6\frac{5}{7} \div 2\frac{7}{9} = \frac{47}{7} \div \frac{25}{9} = \frac{47}{7} \times \frac{9}{25} = \frac{423}{175} = 2\frac{73}{175}$

46. $5\frac{1}{4} \div 2\frac{2}{3} = \frac{21}{4} \div \frac{8}{3} = \frac{21}{4} \times \frac{3}{8} = \frac{63}{32} = 1\frac{31}{32}$

47. $7\frac{1}{2} + 6\frac{3}{4} + 7\frac{7}{8} = \frac{15}{2} + \frac{27}{4} + \frac{63}{8} = \frac{15 \times 4}{2 \times 4} + \frac{27 \times 2}{4 \times 2} + \frac{63}{8}$

$= \frac{60}{8} + \frac{54}{8} + \frac{63}{8} = \frac{177}{8} = 22\frac{1}{8}$ pounds

48. $9\frac{1}{4} - 5\frac{3}{8} = \frac{37}{4} - \frac{43}{8} = \frac{37 \times 2}{4 \times 2} - \frac{43}{8} = \frac{74}{8} - \frac{43}{8} = \frac{31}{8} = 3\frac{7}{8}$ inches

49. $\frac{1}{3} \times \frac{1}{4} = \frac{1}{12}$

50.
$$\begin{array}{r} {}^{1}1.{}^{1}04 \\ + 3.987 \\ \hline 5.027 \end{array}$$

51.
$$\begin{array}{r} {}^{1}7.{}^{1}9{}^{1}09 \\ + {}^{1}16.799 \\ \hline 24.708 \end{array}$$

52.
$$\begin{array}{r} {}^{1}15.{}^{1}3{}^{1}49 \\ + 6.87 \\ \hline 22.219 \end{array}$$

53.
$$\begin{array}{r} {}^{1}1{}^{1}0{}^{1}0.{}^{1}66 \\ + 299.54 \\ \hline 400.20 \end{array}$$

54.
$$\begin{array}{r} {}^{3}4{}^{15}\cancel{6}.{}^{12}\cancel{3}{}^{1}0 \\ - 19.42 \\ \hline 26.88 \end{array}$$

55.
$$\begin{array}{r} {}^{9}\cancel{10}{}^{13}4.{}^{1}1{}^{8}9{}^{1}0 \\ - 55.364 \\ \hline 48.826 \end{array}$$

56.
$$\begin{array}{r} 3.96 \\ \times\, 1.53 \\ \hline 1188 \\ 1\,980 \\ 3\,96 \\ \hline 6.0588 \end{array}$$

58.
$$\begin{array}{r} 10.70 \\ \times\, 19.52 \\ \hline 2140 \\ 5\,350 \\ 96\,30 \\ 107\,0 \\ \hline 208.8640 \end{array}$$

57.
$$\begin{array}{r} 18.56 \\ \times\, 13.08 \\ \hline 1\,4848 \\ 55\,680 \\ 185\,6 \\ \hline 242.7648 \end{array}$$

59. $\frac{1.06}{0.87} = \frac{106}{87}$

$$\begin{array}{r} 1.2 \\ 87\,\overline{)\,106.0} \\ -\,87\,X \\ \hline 19\,0 \\ -\,17\,4 \\ \hline 1\,6 \end{array}$$

60. $\frac{12}{1.16} = \frac{1200}{116} = \frac{300}{29}$

$$\begin{array}{r} 10.3 \\ 29\,\overline{)\,300.0} \\ -\,29X\,X \\ \hline 10\,0 \\ -\,8\,7 \\ \hline 1\,3 \end{array}$$

61. $\frac{90}{4.17} = \frac{9000}{417} = \frac{3000}{139}$

$$\begin{array}{r} 21.5 \quad = 21.6 \\ 139\,\overline{)\,3000.0} \\ -\,278X\,X \\ \hline 220 \\ -\,139 \\ \hline 81\,0 \\ -\,69\,5 \\ \hline 11\,5 \end{array}$$

62. four hundredths $= \frac{4}{100}$ (or $\frac{1}{25}$); 0.04

63. thirty-one thousandths = $\frac{31}{1000}$; 0.031

64. six hundred ninety-one thousandths = $\frac{691}{1000}$; 0.691

65. $\frac{4}{5} = 0.8$

71. $0.3 \times 1,000 = 30$

66. $\frac{9}{15} = \frac{3}{5}$

$$5\overline{)3.0}\quad 0.6$$

72. $1.5 \times 1,000 = 1,500$

67. $\frac{48}{200} = \frac{24}{100} = 0.24$

73. $0.092 \times 1,000 = 92$

68. $0.93 = \frac{93}{100}$

74. $6 \div 100 = 0.06$

69. $0.003 = \frac{3}{1000}$

75. $0.1 \div 100 = 0.001$

70. $0.47 = \frac{47}{100}$

76. $0.004 \div 100 = 0.00004$

77.
$$
\begin{array}{r}
\$1.50 \\
0.90 \\
0.75 \\
+\ 0.08 \\
\hline
\$3.23
\end{array}
$$

78.
$$
\begin{array}{r}
\$8.99 \\
\times\ 16.2 \\
\hline
1\ 798 \\
53\ 94 \\
89\ 9\quad \\
\hline
\$145.638 \\
\end{array}
$$
$= \$145.64$

79.
$$
\begin{array}{r}
\$1.79 \\
\times\ 4.25 \\
\hline
895 \\
358 \\
7\ 16\quad \\
\hline
\$7.6075
\end{array}
\qquad
\begin{array}{r}
\$4.50 \\
\times\ 4.5 \\
\hline
2\ 250 \\
18\ 00\quad \\
\hline
20.250
\end{array}
\qquad
\begin{array}{r}
\$\ 7.6075 \\
+20.25\quad \\
\hline
27.8575 \quad \text{or } \$27.86
\end{array}
$$

80. $0.9 = 90\%$

81. $1.62 = 162\%$

82. $0.07 = 7\%$

83. $20\% = 0.20$ or 0.2

84. $150\% = 1.50$ or 1.5

85. $4\% = 0.04$

86. $\frac{7}{10} = 0.7 = 70\%$

87. $\frac{3}{8} = 0.375 = 37.5\%$

88. $\frac{91}{100} = 91\%$

89. $14\% = \frac{14}{100}$ or $\frac{7}{50}$

90. $200\% = \frac{200}{100} = \frac{2}{1}$ (or 2)

91. $13.9\% = \frac{13.9}{100} = \frac{139}{1000}$

92. $\frac{\$50}{\$400} = \frac{1}{8} = 12.5\%$

93. $\frac{20}{180} = \frac{1}{9} = 11.1\%$

94. $\frac{25}{225} = \frac{1}{9} = 11.1\%$

95.

blue	$= \frac{4}{10}$	$=$	40%
red	$= \frac{3}{10}$	$=$	30%
green	$= \frac{2}{10}$	$=$	20%
yellow	$= \frac{1}{10}$	$=$	10%
Check:			100%

96. $1 - \left(\frac{1}{3} + \frac{2}{5}\right) = 1 - \left(\frac{1 \times 5}{3 \times 5} + \frac{2 \times 3}{5 \times 3}\right) = 1 - \left(\frac{5}{15} + \frac{6}{15}\right) = 1 - \frac{11}{15} = \frac{4}{15} =$

$$
\begin{array}{r}
0.266 \quad = 26.7\% \\
15 \overline{)4.000} \\
\underline{3\ 0XX} \\
1\ 00 \\
\underline{90} \\
100 \\
\underline{90} \\
10
\end{array}
$$

97.

$$
\begin{array}{r}
\$0.06 \\
0.15 \\
0.40 \\
\underline{0.25} \\
\$0.86
\end{array}
$$
$\frac{\$0.86}{\$1.00} = 86\%$

98. $50 + (50 \times 0.2) = 50 + 10 = 60$ mph

99.
$$
\begin{array}{r}
\$16.85 \\
\times\,0.15 \\
\hline
8425 \\
1\,685 \\
\hline
\$2.5275
\end{array}
$$
$\quad = \$2.53$

LAST STEP

You know the drill. I put the lesson numbers on each set of problems so you'd know what lessons to go back and review if you needed further help. If you're doing fine and are ready to go on to more complicated math, turn to the appendix called Additional Resources to see what to tackle next.

ADDITIONAL RESOURCES

Are you ready to tackle algebra, or would you like to work your way through another book like this one to get more practice? Two very similar books are these:

- **Practical Math in 20 Minutes a Day** by Judith Robinovitz (LearningExpress, order information at the back of this book)
- **Arithmetic the Easy Way** by Edward Williams and Katie Prindle (Barron's)

Three other books, which cover much of the same material but also introduce very elementary algebra, as well as some business math applications, are these:

- **Business Mathematics the Easy Way** by Calman Goozner (Barron's)

- **Quick Business Math** by Steve Slavin (Wiley)
- **All the Math You'll Ever Need** by Steve Slavin (Wiley)

Algebra is traditionally taught in a three-year sequence. If you've mastered fractions, decimals, and percentages, then you're definitely ready to tackle elementary (or first-year) algebra.

Unfortunately, many of the algebra books you'll run across assume a prior knowledge of elementary algebra, or rush through it much too quickly. Two of my own books, **Practical Algebra** and **Quick Algebra Review,** do just that.

There are, however, two elementary algebra books that I do recommend:

- **Prealgebra** by Alan Wise and Carol Wise (Harcourt Brace)
- **Let's Review Sequential Mathematics Course 1** by Lawrence S. Leff (Barron's)

Whatever course you follow, just remember that doing math can be fun and exciting. So don't stop now. You'll be amazed at how much further you can go.

Master the Basics... Fast!

Easy to Use & Understand

If you need to improve your basic skills to move ahead either at work or in the classroom, then our LearningExpress books are designed to help anyone master the skills essential for success. It features 20 easy lessons to help build confidence and skill fast. This series includes real world examples—**WHAT YOU REALLY NEED TO SUCCEED.**

All of these books:

- Give quick and easy instruction
- Provides compelling, interactive exercises
- Share practical tips and valuable advise that can be put to use immediately
- Includes extensive lists of resources for continued learning

LEARNINGEXPRESS®
LearnATest.com™